A Beginner's Guide to Signals and Systems

Krishna Kumar Kishor
Department of Electronics and Communication Engineering
Ahalia School of Engineering and Technology
Ahalia Campus
Palakkad - 678557, Kerala, India

ISBN: 9798884391086

Contents

Preface   11

1 Signals   13

1.1   Continuous Time and Discrete Time Signals   13
1.2   Unit Impule / Unit Sample Signal   15
1.3   Unit Step Signal   17
1.4   Sinusoidal Signal   18
1.5   Exponential Signal   19
1.6   Rectangular Signal   20
1.7   Ramp Signal   21
1.8   Sinc Signal   22
1.9   Deterministic and Random Signals   23
1.10 Periodic and Aperiodic Signals   23
1.11 Even and Odd Signals   31
1.12 Energy and Power Signals   34
1.13 Addition, Subtraction, Multiplication of Signals   37
1.14 Multiplying a Signal by a Constant   38
1.15 Signal Operation - Time Shift   39
1.16 Signal Operation - Time Reversal   41
1.17 Signal Operation - Time Scale   42
1.18 Combining Signal Operations   45
1.19 Orthogonal and Orthonormal Signals   47
1.20 Correlation between Signals   50

2 Systems   51

2.1   Representing Systems   51
2.2   Static and Dynamic Systems   53
2.3   Inverse Systems   55
2.4   Causal Systems   57
2.5   Stable Systems   57
2.6   Time-invariant Systems   60
2.7   Linear Systems   62
2.8   Linear Time-invariant (LTI) Systems   64

2.9   Impulse Response Signal 64

2.10 Input-Output relation for DT LTI Systems (Discrete Time Convolution) 65

2.11 Input-Output relation for CT LTI Systems (Continuous Time Convolution) 67

2.12 Commutative Property of the Convolution Operator 68

2.13 Distributive Property of the Convolution Operator 69

2.14 Associative Property of the Convolution Operator 69

2.15 Static and Dynamic Property of LTI Systems 70

2.16 Invertibility of LTI Systems 71

2.17 Causality of LTI Systems 72

2.18 Stability of LTI Systems 73

3 Transforms   **77**

3.1   Continuous Time Fourier Series (CTFS) 77

3.2   Continuous Time Fourier Transform (CTFT) 90

3.3   Laplace Transform (LT) 98

3.4   Discrete Time Fourier Series (DTFS) 111

3.5   Discrete Time Fourier Transform (DTFT) 121

3.6   Z Transform 130

4 Analysing LTI Systems Using Transforms   **147**

4.1   Analysing CT LTI Systems Using CTFT 147

4.2   Analysing CT LTI Systems Using Laplace Transform 151

4.3   Analysing DT LTI Systems Using DTFT 159

4.4   Analysing DT LTI Systems Using Z Transform 161

5 Sampling   **173**

5.1   Derivation of Sampling Theorem 173

5.2   Sampling Theorem 178

5.3   Aliasing 180

Bibliography   **183**

About the Author   **185**

List of Figures

1.1   Concept map of topics in Chapter 1 14
1.2   A continuous time signal 15
1.3   A discrete time signal, $r[n]$ 16
1.4   Unit sample function $\delta[n]$ 16
1.5   Impulse function $\delta(t)$ 16
1.6   DT unit step $u[n]$ 17
1.7   CT unit step $u(t)$ 18
1.8   CT rectangular pulse $x(t)$ 20
1.9   DT rectangular pulse $x[n]$ 20
1.10 Plot of the ramp function, $r(t)$ 21
1.11 Plot of the ramp function, $r[n]$ 22
1.12 Plot of the sinc function 23
1.13 A known signal $cos(t)$ 24
1.14 A random signal 24
1.15 Continuous-time periodic signals 25
1.16 Discrete-time periodic signals 26
1.17 Even and odd signals 32
1.18 A signal $x(t)$ 38
1.19 $z(t) = 2x(t)$ 38
1.20 $w(t) = \frac{1}{2}x(t)$ 39
1.21 $z(t) = x(t-3)$ 40
1.22 $z(t) = x(t+3)$ 41
1.23 $p(t) = x(-t)$ 42
1.24 $p(t) = x(2t)$ 43
1.25 $p(t) = x(\frac{1}{2}t)$ 44
1.26 A signal $x(t)$ 45
1.27 A signal $p(t) = x(3-3t)$ 46
1.28 A signal $p(t) = x(2(t-2))$ 47
1.29 A signal $x(t)$ 48

2.1   Concept map of topics in Chapter 2 52
2.2   Continuous time system 53
2.3   Discrete time system 53

2.4   Example of a static system - resistor   54
2.5   Example of a dynamic system - delay   54
2.6   Inverse system   55
2.7   A bounded signal, $y(t) = 2$ 57
2.8   An unbounded signal, $y(t) = e^{2t}$ 58
2.9   An unbounded signal, $y(t) = t$ 59
2.10 Checking time-invariance - path 1   60
2.11 Checking time-invariance - path 2   60
2.12 Impulse response function   64
2.13 Time-shifted impulse response   65
2.14 e.g. x[n] represented as a sum of time-shifted impulse functions . . .   66
2.15 Commutative property of the convolution operator   68
2.16 Distributive property of the convolution operator   69
2.17 Associate property of the convolution operator   70
2.18 A discrete-time system   71
2.19 Plot of $h(t)$ 73

3.1   Concept map of topics in Chapter 3   78
3.2   CTFS synthesis equation   79
3.3   CTFS analysis equation   80
3.4   Plot of a_k   81
3.5   A periodic signal $x(t)$ (for solved example in Section 3.1.1)   82
3.6   A periodic signal $g(t)$ (for solved example in Section 3.1.3)   89
3.7   CTFT synthesis equation   90
3.8   CTFT analysis equation   91
3.9 $S(j\omega)$ 97
3.10 $P(j\omega)$ 98
3.11 $R(j\omega)$ 99
3.12 Laplace Transform synthesis equation   99
3.13 Laplace Transform analysis equation   100
3.14 ROC: $Re(s) > -a$ 102
3.15 ROC: $Re(s) < -a$ 103
3.16 ROC: $Re(s) > -1$ 107
3.17 ROC: $-b < Re(s) < b$ 109
3.18 DTFS synthesis equation   112
3.19 DTFS analysis equation   113
3.20 A DT periodic square wave   114
3.21 Plot of $x[n]$ 117
3.22 DTFT synthesis equation   121
3.23 DTFT analysis equation   122

3.24 A DT LTI system 128
3.25 Z Transform synthesis equation 131
3.26 Z Transform analysis equation 131
3.27 ROC: $|z| > |a|$ 134
3.28 Plot of $u[-n-1]$ 134
3.29 ROC: $|z| < |a|$ 135
3.30 Pole zero diagram 139
3.31 ROC: $|z| > 1/2$ 143
3.32 ROC: $|z| > 1/3$ 144
3.33 ROC: $1/4 < |z| < 1/3$ 145
3.34 ROC: $|z| < 1/4$ 146

4.1   Concept map of topics in Chapter 4 148
4.2   ROC: $Re(s) > -1$ 154
4.3   ROC: $Re(s) > 2$ 155
4.4   ROC: $Re(s) > 2$ 155
4.5   ROC: $-1 < Re(s) < 2$ 156
4.6   ROC: $Re(s) < -1$ 156
4.7   ROC of $X(s)$: $Re(s) > -3$ 157
4.8   ROC of $Y(s)$: $Re(s) > -1$ 158
4.9   ROC of $H(s)$: $Re(s) > -1$ 158
4.10 ROC: $|z| > 2$ 164
4.11 ROC: $1/2 < |z| < 2$ 165
4.12 ROC: $|z| < 1/2$ 166
4.13 ROC: $1/4 < |z| < 2$ 167
4.14 ROC: $|z| > 2$ 168
4.15 ROC: $|z| > 2$ 169

5.1   Concept map of topics in Chapter 5 173
5.2   A continuous time signal $x(t)$ 174
5.3   Impulse train $p(t)$ 174
5.4   The sampled signal $x_p(t)$ 175
5.5   The sampled signal $x_p(t)$ with lower T (i.e. higher sampling frequency) 176
5.6 $X(j\omega)$ 176
5.7 $P(j\omega)$ 177
5.8 $X_p(j\omega)$ 177
5.9   Frequency response of an ideal low-pass filter 178
5.10 Frequency response of the reconstructed signal (after low-pass filter) 178
5.11 Effect of aliasing 181
5.12 Spectrum $X(f)$ 181

5.13 Spectrum when sampled at 5 kHz 181
5.14 Spectrum when sampled at 3 kHz 182

List of Tables

1.1   Verifying time shift to the right 40

1.2   Verifying time shift to the left 41

1.3   Verifying time reversal 42

1.4   Verifying time scaling for $p(t) = x(2t)$. 44

1.5   Verifying time scaling for $p(t) = x(\frac{1}{2}t)$. 45

2.1   Verifying if the system $y(t) = x(2t)$ is static or dynamic 55

2.2   Verifying if the system $y[n] = nx[n]$ is static or dynamic 55

3.1   The six transforms in this book 77

3.2   Common CTFT pairs 92

3.3   Common Laplace Transform pairs 104

3.4   Common DTFT pairs 125

3.5   Common Z Transform pairs 140

Preface

The concept of a signal and a system is a fundamental topic in the undergraduate curriculum of electrical engineering and its related programs in various universities across the world. These are typically taught in a second year course, as the concepts covered are the foundation (or pre-requisite) of the advanced courses in the fields of communication, signal processing and control systems.

These topics, being mathematically oriented, are often considered difficult by the students. Moreover, the assessments or exams to test their understanding are also problem-based. As such this book is written with a focus on guiding the student on how to solve the numerical problems related to the various concepts presented in this book. For each concept, the relevant theory and several solved examples are presented.

The book is divided into five chapters as shown in the table of contents. For each chapter a concept map is shown, which I hope, helps the students to understand the inter-relation between the various concepts in the chapter. The material presented in this book is out experience of teaching the second year course "Signals and Systems" in Kerala's A. P. J. Abdul Kalam Technological University's syllabus. This introductory book is indended for beginners, particularly undergraduate students, who are interested to learn how to solve numerical problems related to the concepts covered in this book. For those students who are interested to understand and learn more on these concepts, the books mentioned in the reference section [1, 2] will be helpful. For more numerical problems [3, 4, 5] can be referred.

Acknowledgements

I am extremely grateful to Prof. Ravi Adve (at the University of Toronto), for teaching me Signals and Systems in my second year of undergraduate studies. I am also thankful to Prof. Sean Hum and Prof. Frank Kschischang at the University of Toronto. A thanks to all my students who keep me on my toes and have been generous in their comments and suggestions. I also take this opportunity to thank my parents, sister, uncle Dr. V. S. Gopal, teachers, wife and son without whom this book would not be possible.

Feedback

Dear readers, I welcome your suggestions, comments, questions and corrections. Please email me at krishna.k.kishor@ahalia.ac.in.

1 Signals

This chapter discusses about signals in the context of engineering. It aims to answer the following questions:

1. What are the various types of signals? How are signals classified?

2. What are some of the common signals?

3. What are some of the operations on signals? How can signals be manipulated?

4. How do we know the relation between or the degree of similarity between two signals?

The concept map for the topics covered in this chapter is shown in Figure 1.1.

1.1 Continuous Time and Discrete Time Signals

1.1.1 Continuous Time Signals

Figure 1.2 shows an example of a continuous time signal. This is called a continuous time signal since the value of the signal is defined for a continum of values of the independent variable (in this case, time). Thus, those signals where the independent variable is continuous are called continuous time signals.

A continuous time signal is represented as $x(t)$, where the signal is defined by the function x and the independent variable t is time. Note the use of the round brackets (and).

1.1.2 Discrete Time Signals

Figure 1.3 shows an example of a discrete time signal. This is called a discrete time signal since the value of the signal is defined for only for discrete instances of the independent variable (in this case, time). Thus, those signals where the independent variable takes on a discrete set of values are called discrete time signals.

A discrete time signal is represented as $x[n]$, where the signal is defined by the function x and the independent variable n is time. Note that n only takes on integer values $(0, \pm1, \pm2, \ldots)$. Also note the use of the square brackets [and]. Also note

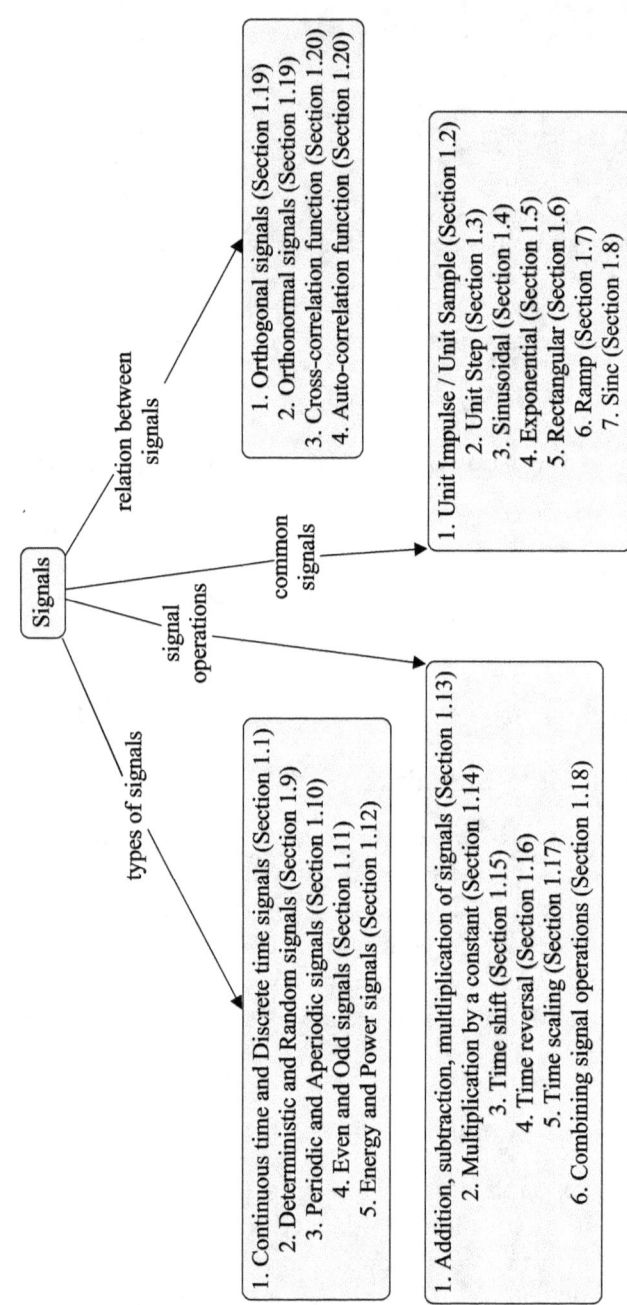

Figure 1.1: Concept map of topics in Chapter 1

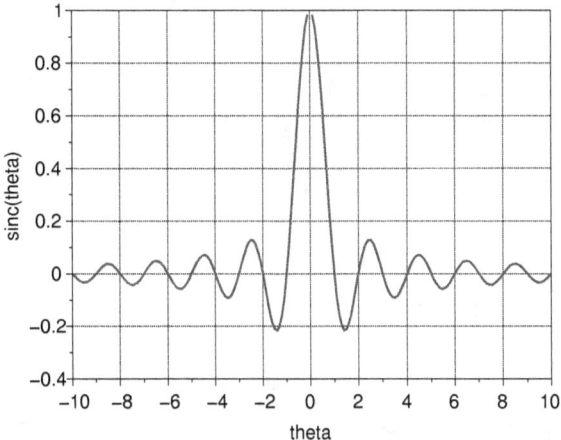

Figure 1.2: A continuous time signal

that in between the samples, we do not know what the value of the signal is. That is, at $n = 3.5$ a discrete time signal is undefined.

1.2 Unit Impule / Unit Sample Signal

1.2.1 Unit Sample Signal in Discrete Time

Figure 1.4 shows the unit sample signal. It is denoted by $\delta[n]$. It is defined as:

$$\delta[n] = \begin{cases} 1, & n = 0 \\ 0, & n \neq 0 \end{cases} \tag{1.1}$$

An important property of $\delta[n]$, that is often used in numerical problems, is the sampling property of $\delta[n]$. It is given by:

$$x[n]\delta[n - n_0] = x[n_0]\delta[n - n_0] \tag{1.2}$$

1.2.2 Unit Impulse Signal in Continuous Time

Figure 1.5 shows the unit impulse signal. It is denoted by $\delta(t)$. It is defined as:

$$\int_{-\infty}^{\infty} \delta(t)dt = 1 \tag{1.3}$$

15

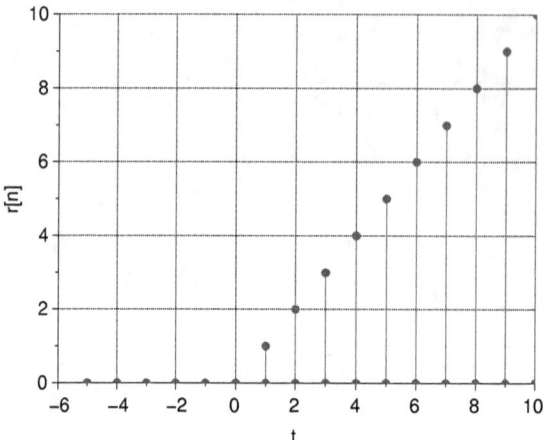

Figure 1.3: A discrete time signal, $r[n]$

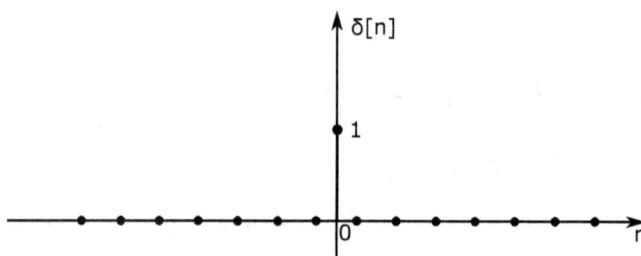

Figure 1.4: Unit sample function $\delta[n]$

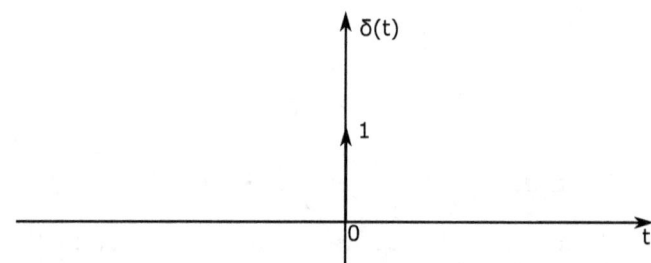

Figure 1.5: Impulse function $\delta(t)$

An important property of $\delta(t)$ is the sampling property of $\delta(t)$. It is given by:

$$x(t)\delta(t - t_0) = x(t_0)\delta(t - t_0). \tag{1.4}$$

Integrating both sides of the above equation, we get another important result.

$$\int_{-\infty}^{\infty} x(t)\delta(t - t_0)dt = x(t_0) \ \left(\text{since } \int_{-\infty}^{\infty} \delta(t - t_0)dt = 1\right) \tag{1.5}$$

1.3 Unit Step Signal

1.3.1 Unit Step Signal in Discrete Time

Figure 1.6 shows the unit step signal. It is denoted by $u[n]$. It is defined as:

$$u[n] = \begin{cases} 0, & n < 0 \\ 1, & n \geq 0 \end{cases} \tag{1.6}$$

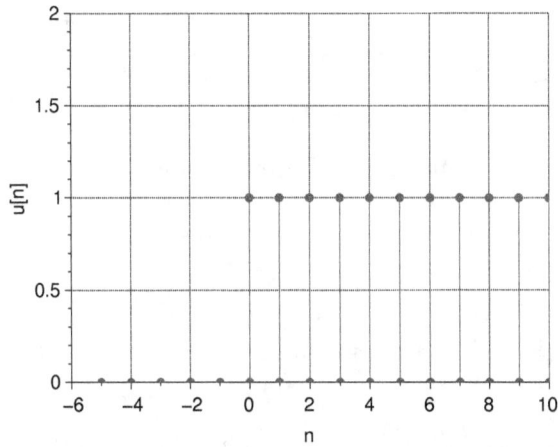

Figure 1.6: DT unit step $u[n]$

The unit sample signal ($\delta[n]$) and the unit step signal ($u[n]$) are related through the following expressions:

$$\delta[n] = u[n] - u[n - 1] \tag{1.7}$$

$$u[n] = \sum_{m=-\infty}^{n} \delta[m] \tag{1.8}$$

$$u[n] = \sum_{k=0}^{\infty} \delta[n - k] \tag{1.9}$$

17

1.3.2 Unit Step Signal in Continuous Time

Figure 1.7 shows the unit step signal. It is denoted by $u(t)$. It is defined as:

$$u(t) = \begin{cases} 0, & t < 0 \\ 1, & t \geq 0 \end{cases} \tag{1.10}$$

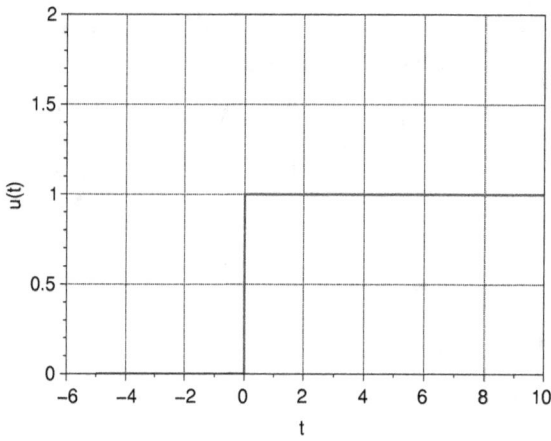

Figure 1.7: CT unit step $u(t)$

The unit impulse signal $(\delta(t))$ and the unit step signal $(u(t))$ are related through the following expressions:

$$u(t) = \int_{\tau=-\infty}^{t} \delta(\tau)d\tau \text{ (similar to Equation (1.8))} \tag{1.11}$$

$$\Rightarrow \delta(t) = \frac{du(t)}{dt} \text{ (similar to Equation (1.7))} \tag{1.12}$$

$$u(t) = \int_{\sigma=0}^{\infty} \delta(t-\sigma)d\sigma \text{ (similar to Equation (1.9))} \tag{1.13}$$

1.4 Sinusoidal Signal

1.4.1 Sinusoidal Signal in Continuous Time

The well-known sinusoidal signal is defined as:

$$x(t) = A\sin(\omega t + \phi) \tag{1.14}$$

18

where ω and ϕ are the angular frequency in radians per second (rad/s) and the phase shift in radians respectively. ω and the frequency (f) in Hertz (Hz) are related via $\omega = 2\pi f$. Note that $x(t)$ is a periodic signal with time period T seconds, where T can be any positive real number. f and T are related via $f = 1/T$. Since T can be any positive real number, ω and T are related via $T = 2\pi/\omega$.

1.4.2 Sinusoidal Signal in Discrete Time

The sinusoidal signal in discrete time is defined as:

$$x[n] = A\cos(\Omega n + \phi) \tag{1.15}$$

where Ω and ϕ are the angular frequency in radians per second and the phase shift in radians respectively. Note that $x[n]$ is a perioid signal with time period N seconds, only if N is a positive integer. Thus to derive the relation between N and Ω we rely on the definition of the period signal seen in Section 1.10. That is, for $x[n]$ to be perioidic in N, $x[n] = x[n + N]$. Therefore,

$$x[n + N] = A\cos(\Omega(n + N) + \phi) = A\cos(\Omega n + \Omega N + \phi). \tag{1.16}$$

Comparing Equations (1.15) and (1.16), if $x[n + N] = x[n]$, then

$$\Omega N = m(2\pi), \text{where } m \text{ is an integer}$$

$$\Omega = \frac{m}{N}(2\pi) \tag{1.17}$$

$$N = m\frac{2\pi}{\Omega} \tag{1.18}$$

Equation (1.17) suggests that Ω must be a rational multiple of 2π. This is the main difference between Ω (of discrete time) and ω (of continuous time). Equivalently, Equation (1.18) suggests that for $x[n]$ to be periodic there must exist an integer m such that N is an integer.

1.5 Exponential Signal

1.5.1 Exponential Signal in Continuous Time

An exponential signal is defined as:

$$x(t) = Ae^{bt} \tag{1.19}$$

where A is assumed to be real, b can be a real number (in which case $x(t)$ is a real signal) or a complex number (in which case $x(t)$ is a complex valued signal). Note that:

- If b is real, b can be positive ($b > 0$, in which case the signal is exponentially increasing) or negative ($b > 0$, in which case the signal is exponentially decreasing).

- If b is complex, b can be purely imaginary ($b = j\omega_0$, in which case the magnitude of the signal is a constant) or can have a real and imaginary parts ($b = \sigma_0 + j\omega_0$).

1.5.2 Exponential Signal in Discrete Time

An exponential signal is defined as:

$$x[n] = Cz^n \tag{1.20}$$

where C and z are in general complex numbers.

1.6 Rectangular Signal

Figures 1.8 and 1.9 show a rectangular signal in continuous time and discrete time respectively.

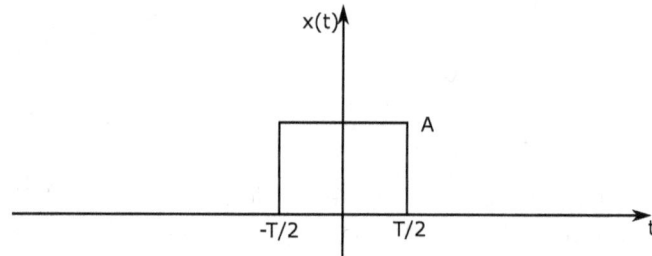

Figure 1.8: CT rectangular pulse $x(t)$

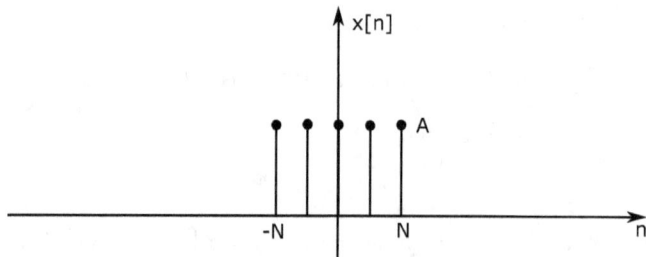

Figure 1.9: DT rectangular pulse $x[n]$

In continuous time, the rectangular signal (or *rect*() function) is defined as:

$$x(t) = \begin{cases} A, & -T/2 \leq t \leq T/2 \\ 0, & \text{otherwise} \end{cases} \tag{1.21}$$

A similar definition can be given for the rectangular signal in discrete time.

1.7 Ramp Signal

1.7.1 Ramp Signal in Continuous Time

A ramp signal in continuous time is defined as:

$$r(t) = \begin{cases} t, & t \geq 0 \\ 0, & t < 0 \end{cases} \tag{1.22}$$

Equivalently, it can also be defined as:

$$r(t) = tu(t). \tag{1.23}$$

A plot of $r(t)$ is shown in Figure 1.10

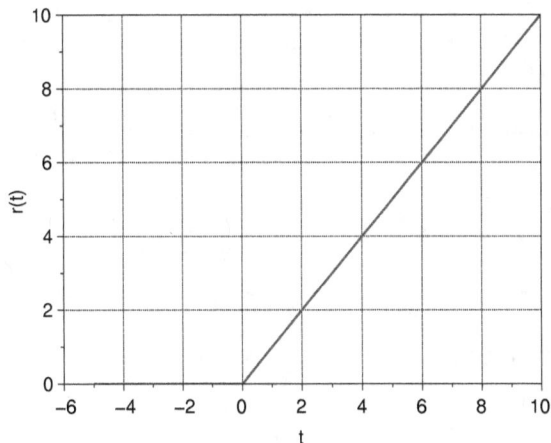

Figure 1.10: Plot of the ramp function, $r(t)$

21

1.7.2 Ramp Signal in Discrete Time

A ramp signal in discrete time is defined as:

$$r[n] = \begin{cases} n, & n \geq 0 \\ 0, & n < 0 \end{cases} \tag{1.24}$$

Equivalently, it can also be defined as:

$$r[n] = nu[n] \tag{1.25}$$

A plot of $r[n]$ is shown in Figure 1.11

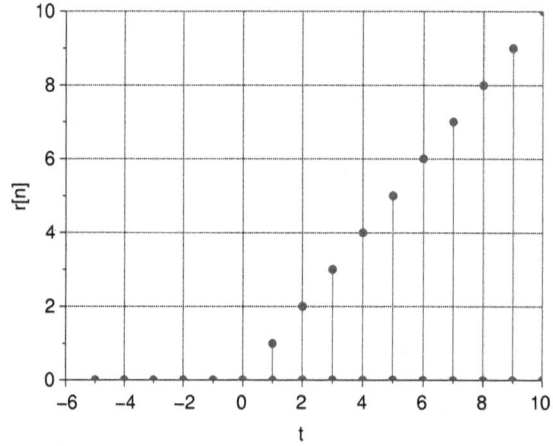

Figure 1.11: Plot of the ramp function, $r[n]$

1.8 Sinc Signal

The $sinc()$ signal in continuous time is defined as:

$$\text{sinc}(\theta) = \frac{\sin(\pi\theta)}{\pi\theta} \tag{1.26}$$

A plot of the sinc function is shown in Figure 1.12.

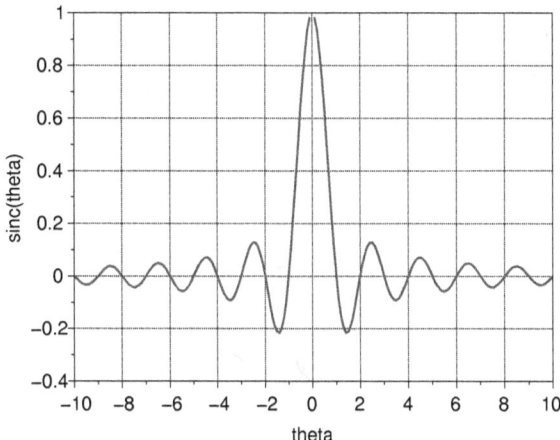

Figure 1.12: Plot of the sinc function

1.9 Deterministic and Random Signals

1.9.1 Deterministic Signals

Those signals whose value at any instant of time is known are called deterministic signals. $x(t) = \cos(t)$ shown in Figure 1.13 is an example of a deterministic signal.

1.9.2 Random Signals

Those signals about which there is an uncertainty in its value are called random signals. Figure 1.14 is an example of a random signal. Practical examples include noise signals in an electrical circuit and signals received at the receiver antenna after propagation in a wireless channel.

1.10 Periodic and Aperiodic Signals

1.10.1 Periodic Signals

A continuous-time signal $x(t)$ is said to be periodic if it satisfies the following relation

$$x(t) = x(t + T), \tag{1.27}$$

where T is a positive real number. i.e. $x(t)$ is periodic with T. It can be shown that if $x(t) = x(t + T)$, then $x(t) = x(t + mT)$, where m is any integer. i.e. $x(t)$ is periodic with $T, 2T, 3T, \ldots$.

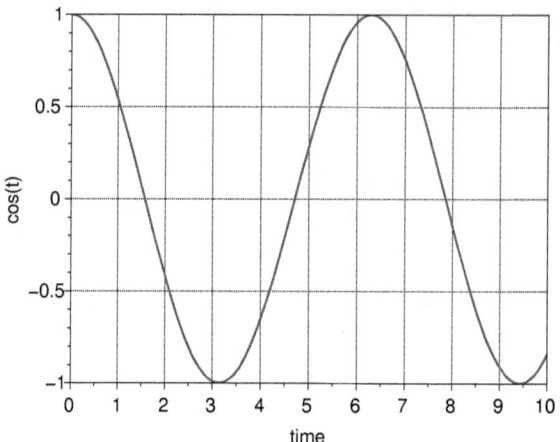

Figure 1.13: A known signal $cos(t)$

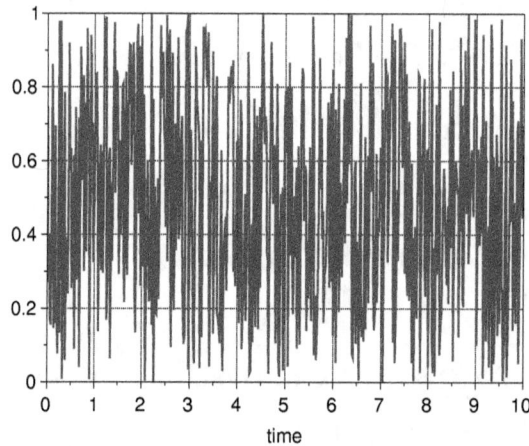

Figure 1.14: A random signal

The fundamental period T_0 of $x(t)$ is the smallest value of T for which (1.27) holds. Figure 1.15 shows three continuous-time periodic signals at different frequencies.

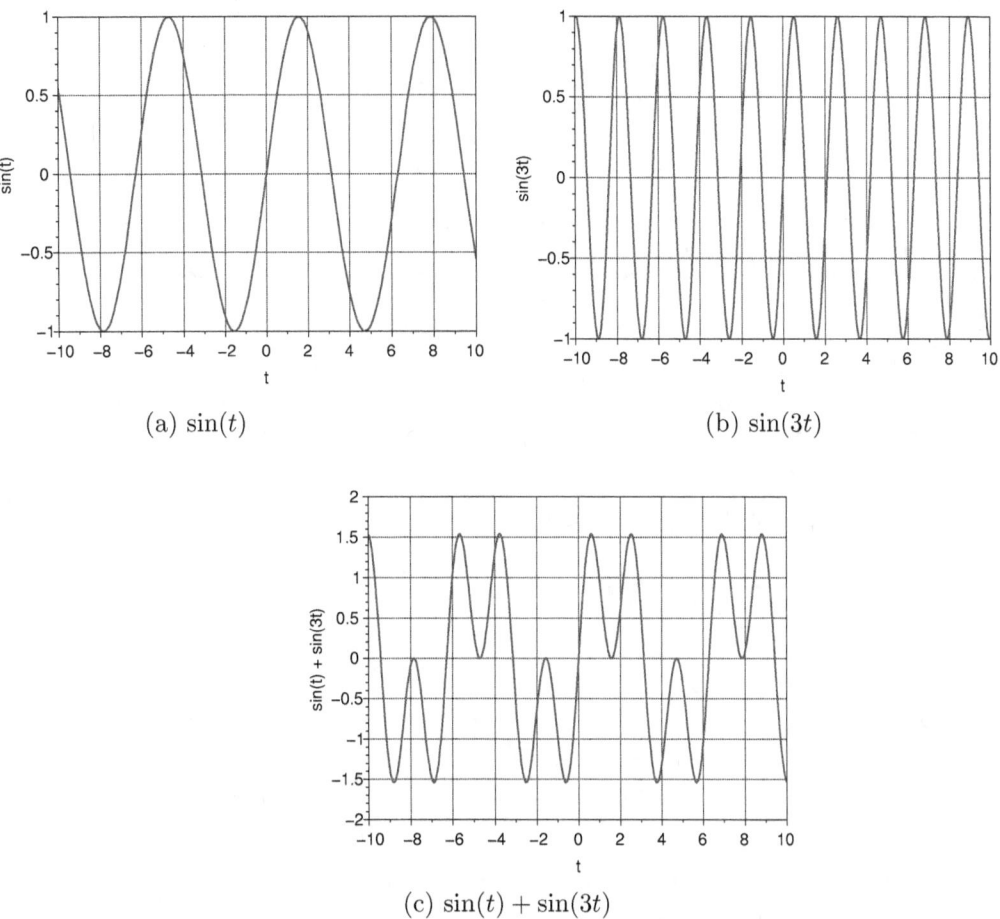

(a) $\sin(t)$

(b) $\sin(3t)$

(c) $\sin(t) + \sin(3t)$

Figure 1.15: Continuous-time periodic signals

A discrete-time signal $x[n]$ is said to be periodic if it satisfies the following relation

$$x[n] = x[n+N], \tag{1.28}$$

where N is a positive integer value. i.e. $x[n]$ is periodic with N.

The fundamental period N_0 of $x[n]$ is the smallest value of N for which (1.28) holds. Figure 1.16 shows three discrete-time periodic signals at different frequencies.

25

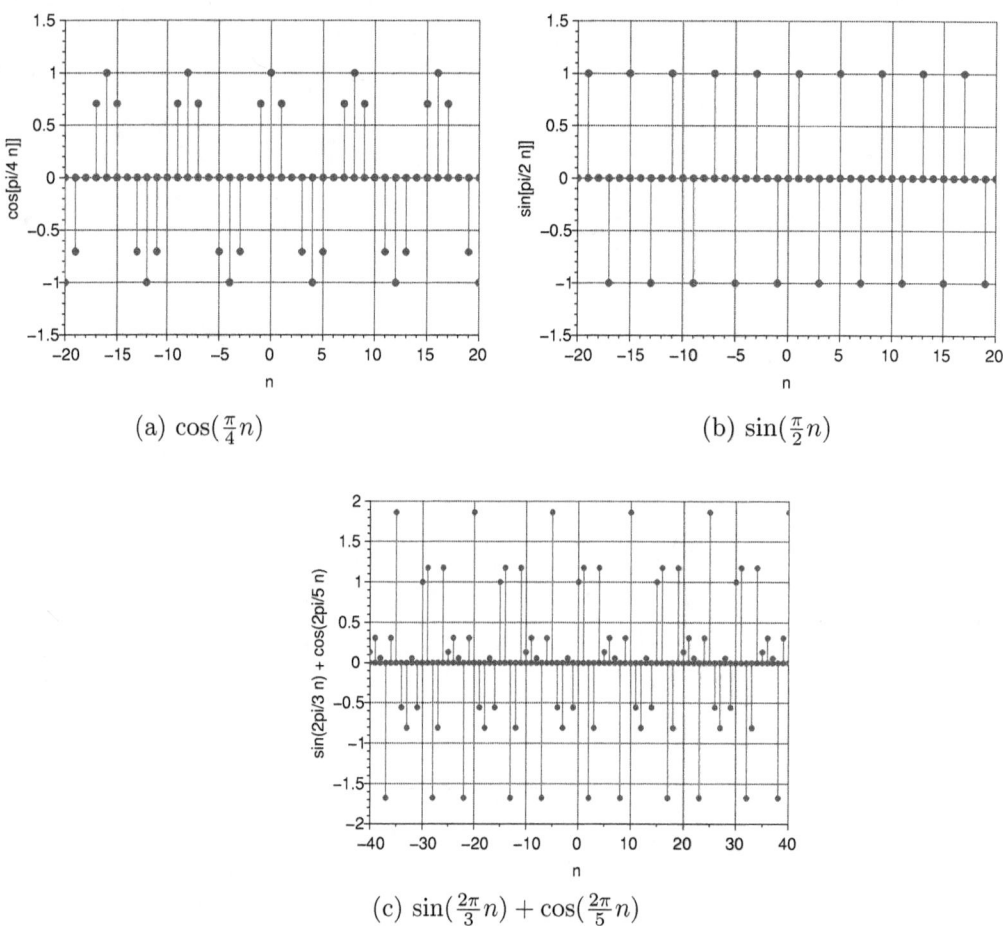

(a) $\cos(\frac{\pi}{4}n)$

(b) $\sin(\frac{\pi}{2}n)$

(c) $\sin(\frac{2\pi}{3}n) + \cos(\frac{2\pi}{5}n)$

Figure 1.16: Discrete-time periodic signals

1.10.2 Aperiodic Signals

Signals that are not periodic are called aperiodic signals; that is, they do not satisfy Equations (1.27) or (1.28). Figure 1.14 is an example of a continuous-time aperiodic signal. Figure 1.9 is an example of a discrete-time aperiodic signal.

1.10.3 Solved Examples

> 1. Determine if the signal $x(t) = \sin(t)$ is periodic. If periodic, determine the fundamental period.

Solution: Comparing $x(t)$ with Equation (1.14) we get $\omega = 1$. Thus $T = \frac{2\pi}{\omega} = 2\pi$ [sec] (a real number). Thus $x(t)$ is periodic. A plot of $x(t)$ is shown in Figure 1.15a.

> 2. Determine if the signal $y(t) = \sin(3t)$ is periodic. If periodic, determine the fundamental period.

Solution: Comparing $y(t)$ with Equation (1.14) we get $\omega = 3$. Thus $T = \frac{2\pi}{\omega} = \frac{2\pi}{3}$ [sec] (a real number). Thus $y(t)$ is periodic. A plot of $y(t)$ is shown in Figure 1.15b.

> 3. Is the signal $z(t) = 10\cos^2(10\pi t)$ a periodic signal? If yes, what is its fundamental period?

Solution:

$$z(t) = 10\cos^2(10\pi t)$$
$$= 10\left(\frac{1 + \cos(20\pi t)}{2}\right)$$
$$= 5 + 5\cos(20\pi t)$$

Comparing with Equation (1.14) we get $\omega = 20\pi$ [1]. Thus $T = \frac{2\pi}{\omega} = \frac{2\pi}{20\pi} = \frac{1}{10} = 0.1$ [sec] (a real number). Thus $z(t)$ is periodic.

---

[1]Note that the addition of 5 only results in an amplitude shift and does not affect the periodicity of the signal

4. Determine if the signal $p(t) = \sin(t) + \sin(3t)$ is periodic. If periodic, determine the fundamental period.

Solution: To solve this problem, we utilize the following information: if $x(t)$ and $y(t)$ are two periodic signals with fundamental period T_x and T_y respectively, then $z(t) = x(t) + y(t)$ is also a periodic signal if T_x/T_y is a rational number. The fundamental period T of $z(t)$ is LCM(T_x, T_y).

In this problem $x(t) = \sin(t)$ and $y(t) = \sin(3t)$. From the earlier two examples, the time period of $\sin(t)$ is $T_1 = 2\pi$ and the time period of $\sin(3t)$ is $T_2 = \frac{2\pi}{3}$. Therefore:

$$\frac{T_1}{T_2} = \frac{2\pi}{\frac{2\pi}{3}} = 3.$$

Since $\frac{T_1}{T_2}$ is a rational number, the signal $p(t)$ is periodic. The periodicity is given by LCM$(2\pi, \frac{2\pi}{3}) = 2\pi$ [sec]. A plot of $p(t)$ is shown in Figure 1.15c.

5. Determine if the signal $w(t) = a\sin(4t) + b\cos(7t)$ is periodic. If periodic, determine the fundamental period.

Solution:

$$a\sin(4t) \Rightarrow \omega_1 = 4$$
$$T_1 = \frac{2\pi}{\omega_1} = \frac{\pi}{2}$$

$$b\cos(7t) \Rightarrow \omega_2 = 7$$
$$T_2 = \frac{2\pi}{\omega_2} = \frac{2\pi}{7}$$

Therefore,

$$\frac{T_1}{T_2} = \frac{\frac{\pi}{2}}{\frac{2\pi}{7}} = 7/4.$$

Since $\frac{T_1}{T_2}$ is a rational number, the signal $w(t)$ is periodic. The periodicity is given by LCM$(\frac{\pi}{2}, \frac{2\pi}{7}) = 2\pi$ [sec].

6. Determine if the signal $x(t) = \cos(t) + \sin(\sqrt{2}\, t)$ is periodic. If periodic, determine the fundamental period.

Solution:

$$\cos(t) \Rightarrow \omega_1 = 1$$
$$T_1 = \frac{2\pi}{\omega_1} = 2\pi$$

$$\sin(\sqrt{2}\, t) \Rightarrow \omega_2 = \sqrt{2}$$
$$T_2 = \frac{2\pi}{\omega_2} = \frac{2\pi}{\sqrt{2}} = \sqrt{2}\pi$$

Therefore,
$$\frac{T_1}{T_2} = \frac{2\pi}{\sqrt{2}\pi} = \sqrt{2}.$$

Since $\frac{T_1}{T_2}$ is an irrational number, the signal $x(t)$ is not periodic.

7. Determine if the signal $x[n] = \cos(\frac{\pi}{4}n)$ is periodic. If periodic, determine the fundamental period.

Solution: Comparing $x[n]$ with Equation (1.15) we get $\Omega = \frac{\pi}{4}$. Thus from Equation (1.18),
$$N = m\frac{2\pi}{\Omega} = m\frac{2\pi}{\frac{\pi}{4}} = 8m.$$

For $m = 1$, the fundamental period $N = 8$ (an integer number). Thus $x[n]$ is periodic. A plot of $x[n]$ is shown in Figure 1.16a.

8. Determine if the signal $y[n] = \sin(\frac{\pi}{2}n)$ is periodic. If periodic, determine the fundamental period.

Solution: Comparing $y[n]$ with Equation (1.15) we get $\Omega = \frac{\pi}{2}$. Thus from Equation (1.18),
$$N = m\frac{2\pi}{\Omega} = m\frac{2\pi}{\frac{\pi}{2}} = 4m.$$

For $m = 1$, the fundamental period $N = 4$ (an integer number). Thus $y[n]$ is periodic. A plot of $y[n]$ is shown in Figure 1.16b.

9. Determine if the signal $p[n] = \sin(n)$ is periodic. If periodic, determine the fundamental period.

Solution: Comparing $p[n]$ with Equation (1.15) we get $\Omega = 1$. Thus from Equation (1.18),

$$N = m\frac{2\pi}{\Omega} = m\frac{2\pi}{1} = m(2\pi).$$

For no value of m can we get N to be an integer. Thus $p[n]$ is not periodic.

10. Determine if the signal $w[n] = \cos(4n)$ is periodic. If periodic, determine the fundamental period.

Solution: Comparing $w[n]$ with Equation (1.15) we get $\Omega = 4$. Thus from Equation (1.18),

$$N = m\frac{2\pi}{\Omega} = m\frac{2\pi}{4} = m(\frac{\pi}{2}).$$

For no value of m can we get N to be an integer. Thus $w[n]$ is not periodic.

11. Determine if the signal $x[n] = \cos^2(\frac{\pi}{8}n)$ is periodic. If periodic, determine the fundamental period.

Solution:

$$x[n] = \cos^2(\frac{\pi}{8}n)$$
$$= \frac{1}{2} + \frac{1}{2}\cos(2\frac{\pi}{8}n)$$
$$= \frac{1}{2} + \frac{1}{2}\cos(\frac{\pi}{4}n)$$

Comparing with Equation (1.15) we get $\omega = \frac{\pi}{4}$ [2]. Thus

$$N = m\frac{2\pi}{\Omega} = m\frac{2\pi}{\frac{\pi}{4}} = 8m.$$

---

[2]Note that the addition of 1/2 only results in an amplitude shift and does not affect the periodicity of the signal

For $m = 1$, the fundamental period $N = 8$ (an integer number). Thus $x[n]$ is periodic.

---

2. Determine if the signal $p[n] = \sin(\frac{2\pi}{3}n) + \cos(\frac{2\pi}{5}n)$ is periodic. If periodic, determine the fundamental period.

Solution: To solve this problem, we utilize the following information: if $x[n]$ and $y[n]$ are two periodic signals with fundamental period N_x and N_y respectively, then $z[n] = x[n] + y[n]$ is also a periodic signal if N_x/N_y is a rational number. The fundamental period N of $z[n]$ is LCM(N_x, N_y).

In this problem $x[n] = \sin(\frac{2\pi}{3}n)$ and $y[n] = \cos(\frac{2\pi}{5}n)$.

$$\sin(\frac{2\pi}{3}n) \Rightarrow \Omega = \frac{2\pi}{3}$$
$$N = m\frac{2\pi}{\Omega} = m\frac{2\pi}{\frac{2\pi}{3}} = 3m.$$

For $m = 1$, the fundamental period $N_1 = 3$ (an integer number). Thus $x[n]$ is periodic.

$$\cos(\frac{2\pi}{5}n) \Rightarrow \Omega = \frac{2\pi}{5}$$
$$N = m\frac{2\pi}{\Omega} = m\frac{2\pi}{\frac{2\pi}{5}} = 5m.$$

For $m = 1$, the fundamental period $N_2 = 5$ (an integer number). Thus $y[n]$ is periodic.

Therefore,
$$\frac{N_1}{N_2} = \frac{3}{5}.$$

Since $\frac{N_1}{N_2}$ is a rational number, the signal $p[n]$ is periodic. The periodicity is given by LCM(3,5) = 15.

1.11 Even and Odd Signals

A signal $x(t)$ or $x[n]$ is an even signal if

$$x(-t) = x(t) \tag{1.29}$$
$$x[-n] = x[n] \tag{1.30}$$

31

A signal $x(t)$ or $x[n]$ is an odd signal if

$$x(-t) = -x(t) \tag{1.31}$$
$$x[-n] = -x[n] \tag{1.32}$$

Figures 1.17 shows a continuous-time even and odd signal.

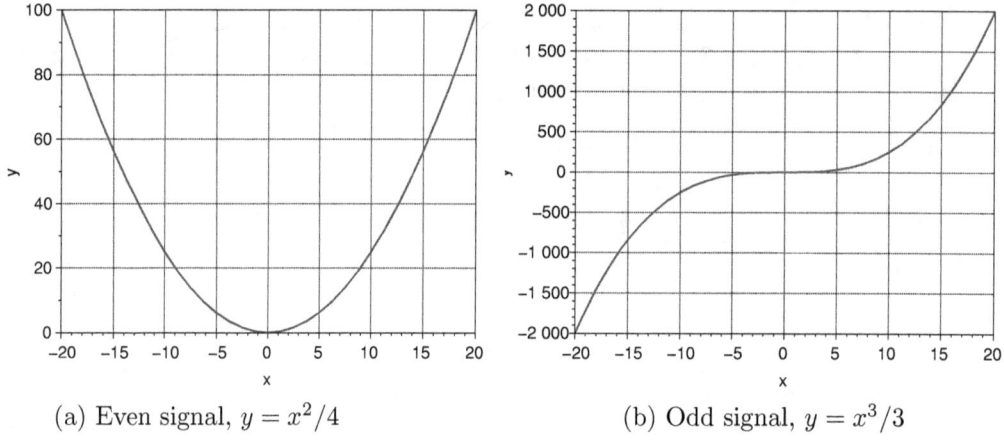

(a) Even signal, $y = x^2/4$ (b) Odd signal, $y = x^3/3$

Figure 1.17: Even and odd signals

Any continuous-time or discrete-time signal can be broken into a sum of two signals; i.e. as a sum of its even and odd components. That is,

$$x(t) = x_e(t) + x_o(t) \tag{1.33}$$
$$x[n] = x_e[n] + x_o[n] \tag{1.34}$$

where $x_e(t)$ or $x_e[n]$ is the even component of $x(t)$ or $x[n]$; and $x_o(t)$ or $x_o[n]$ is the odd component of $x(t)$ or $x[n]$.

The even component $x_e(t)$ or $x_e[n]$ is calculated as:

$$x_e(t) = \frac{x(t) + x(-t)}{2} \tag{1.35}$$
$$x_e[n] = \frac{x[n] + x[-n]}{2}. \tag{1.36}$$

The odd component $x_o(t)$ or $x_o[n]$ is calculated as:

$$x_o(t) = \frac{x(t) - x(-t)}{2} \tag{1.37}$$
$$x_o[n] = \frac{x[n] - x[-n]}{2}. \tag{1.38}$$

1.11.1 Solved Examples

1. Is the following signal odd or even?

$$x(t) = \begin{cases} \sin(\frac{\pi t}{T}), & -T \le t \le T \\ 0, & \text{otherwise} \end{cases}$$

Solution:

$$x(t) = \sin(\frac{\pi t}{T})$$

$$x(-t) = \sin(\frac{-\pi t}{T})$$

$$= -\sin(\frac{\pi t}{T})$$

$$= -x(t)$$

Hence the signal $x(t)$ is an odd signal.

2. Find the even and odd components of $x(t) = e^{-2t}\cos(t)$.

Solution:

$$x_e(t) = \frac{x(t) + x(-t)}{2}$$

$$= \frac{e^{-2t}\cos(t) + e^{2t}\cos(t)}{2}$$

$$= \cos(t)\frac{e^{-2t} + e^{2t}}{2}$$

$$= \cos(t)\cosh(2t).$$

$$x_o(t) = \frac{x(t) - x(-t)}{2}$$

$$= \frac{e^{-2t}\cos(t) - e^{2t}\cos(t)}{2}$$

$$= \cos(t)\frac{e^{-2t} - e^{2t}}{2}$$

$$= -\cos(t)\sinh(2t).$$

1.12 Energy and Power Signals

Before understanding the concepts of energy and power signals, we must know how to calculate the total energy and average power of continuous-time and discrete-time signals.

1.12.1 Energy of a Signal

The total energy of a continuous-time signal $x(t)$ is given by:

$$E = \lim_{T \to \infty} \int_{-T}^{T} |x(t)|^2 dt \tag{1.39}$$

The total energy of a discrete-time signal $x[n]$ is given by:

$$E = \lim_{N \to \infty} \sum_{-N}^{N} |x[n]|^2 \tag{1.40}$$

1.12.2 Average Power of a Signal

The average power of a continuous-time signal $x(t)$ is given by:

$$P_{av} = \lim_{T \to \infty} \frac{1}{2T} \int_{-T}^{T} |x(t)|^2 dt \tag{1.41}$$

The average power of a discrete-time signal $x[n]$ is given by:

$$P_{av} = \lim_{N \to \infty} \frac{1}{2N+1} \sum_{-N}^{N} |x[n]|^2 \tag{1.42}$$

1.12.3 Energy Signal

An energy signal is one whose total energy is finite and non-zero. That is $0 < E < \infty$.

1.12.4 Power Signal

A power signal is one whose average power is finite and non-zero. That is $0 < P_{av} < \infty$.

1.12.5 Solved Examples

1. Determine if the signal

$$x(t) = \begin{cases} A, & 0 \le t \le T_0 \\ 0, & \text{otherwise} \end{cases}$$

is an energy signal or power signal?

Solution: Using Equation (1.39),

$$\begin{aligned}
E &= \lim_{T \to \infty} \int_{-T}^{T} |x(t)|^2 dt \\
&= \int_{-\infty}^{0} |x(t)|^2 dt + \int_{0}^{T_0} |x(t)|^2 dt + \int_{T_0}^{\infty} |x(t)|^2 dt \\
&= 0 + \int_{0}^{T_0} |x(t)|^2 dt + 0 \\
&= \int_{0}^{T_0} A^2 dt \\
&= A^2 \int_{0}^{T_0} dt \\
&= A^2 T_0.
\end{aligned}$$

Since E is finite the signal is an energy signal.

2. Determine if the signal

$$x[n] = \begin{cases} \cos(\pi n), & n \ge 0 \\ 0, & \text{otherwise} \end{cases}$$

is an energy signal or power signal?

Solution: Using Equation (1.40),

$$E = \lim_{N \to \infty} \sum_{-N}^{N} |x[n]|^2$$

$$= \lim_{N \to \infty} \sum_{0}^{N} |x[n]|^2$$

$$= \lim_{N \to \infty} \sum_{0}^{N} \cos^2(\pi n)$$

$$= \lim_{N \to \infty} \sum_{0}^{N} \frac{1 + \cos(2\pi n)}{2}$$

$$= \lim_{N \to \infty} \sum_{0}^{N} \left[\frac{1}{2} + \frac{1}{2} \cos(2\pi n) \right]$$

$$= \lim_{N \to \infty} \sum_{0}^{N} \frac{1}{2} + \lim_{N \to \infty} \sum_{0}^{N} \frac{1}{2} \cos(2\pi n)$$

$$= A + B$$

$$A = \lim_{N \to \infty} \sum_{0}^{N} \frac{1}{2}$$

$$= \lim_{N \to \infty} \frac{1}{2} + \frac{1}{2} + \cdots + \frac{1}{2} \ (N + 1 \text{ times})$$

$$= \lim_{N \to \infty} \frac{1}{2}(N + 1)$$

$$= \infty (\text{as } N \to \infty)$$

Thus $E \to \infty$. Therefore E is not an energy signal. To check whether the signal is a power signal, we use Equation (1.42). Following the steps done earlier, we get:

$$P_{av} = \lim_{N \to \infty} \frac{1}{2N + 1} \sum_{-N}^{N} |x[n]|^2$$

$$= \lim_{N \to \infty} \frac{1}{2N + 1} \sum_{0}^{N} \frac{1}{2} + \lim_{N \to \infty} \frac{1}{2N + 1} \sum_{0}^{N} \frac{1}{2} \cos(2\pi n)$$

$$= A + B$$

$$A = \lim_{N \to \infty} \frac{1}{2N+1} \sum_{0}^{N} \frac{1}{2}$$

$$= \lim_{N \to \infty} \frac{1}{2N+1} \frac{1}{2}(N+1).$$

$$B = \lim_{N \to \infty} \frac{1}{2N+1} \sum_{0}^{N} \frac{1}{2} \cos(2\pi n)$$

$$= \lim_{N \to \infty} \frac{1}{2N+1} \frac{1}{2}[1 + 1 + \cdots + 1 \ (N+1 \ \text{times})]$$

$$= \lim_{N \to \infty} \frac{1}{2N+1} \frac{1}{2}(N+1).$$

Thus

$$P_{av} = \lim_{N \to \infty} \frac{1}{2N+1} \frac{1}{2}(N+1) \times 2$$

$$= \lim_{N \to \infty} \frac{N+1}{2N+1}$$

$$= \lim_{N \to \infty} \frac{1 + \frac{1}{N}}{2 + \frac{1}{N}}$$

$$= \frac{1}{2}.$$

Since P_{av} is a finite value, the signal is a power signal.

1.13 Addition, Subtraction, Multiplication of Signals

If $x(t), y(t)$ are two continuous time signals, then we can also calculate the sum, difference and product of these signals such that:

$$z(t) = x(t) \pm y(t) \tag{1.43}$$

$$w(t) = x(t) \times y(t). \tag{1.44}$$

Note that the operations are done at each time instant. These operations are also defined for discrete time signals $x[n], y[n], z[n]$.

1.14 Multiplying a Signal by a Constant

If $x(t)$ is a CT signal, then we can define a signal:

$$z(t) = ax(t), \qquad\qquad (1.45)$$

where a is a constant. If $|a| > 1$, then $z(t)$ is amplified compared to $x(t)$. If $|a| < 1$, then $z(t)$ is attenuated compared to $x(t)$.

Note that these operations are also defined for discrete time signals $x[n], z[n]$.

1.14.1 Solved Examples

1. Consider a signal given in Figure 1.18. Plot $z(t) = 2x(t)$.

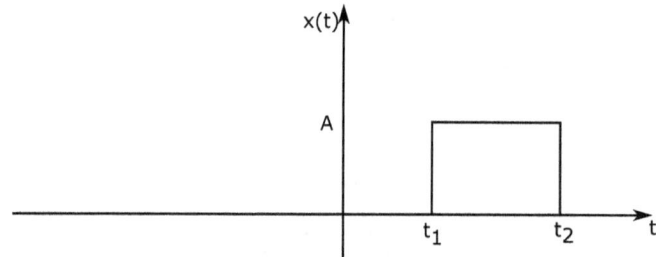

Figure 1.18: A signal $x(t)$

Solution: The amplitude of the signal $z(t)$, at every time instant, is twice the amplitude of the original signal $x(t)$, as shown in Figure 1.19.

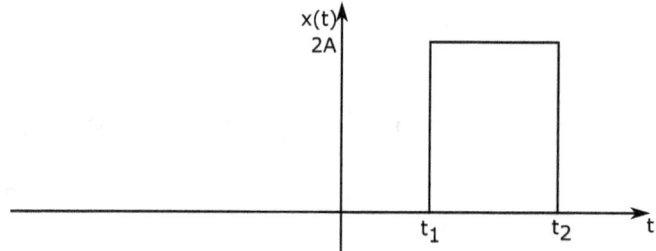

Figure 1.19: $z(t) = 2x(t)$

2. Consider a signal given in Figure 1.18. Plot $w(t) = \frac{1}{2}x(t)$.

Solution: The amplitude of the signal $w(t)$, at every time instant, is half the amplitude of the original signal $x(t)$, as shown in Figure 1.20.

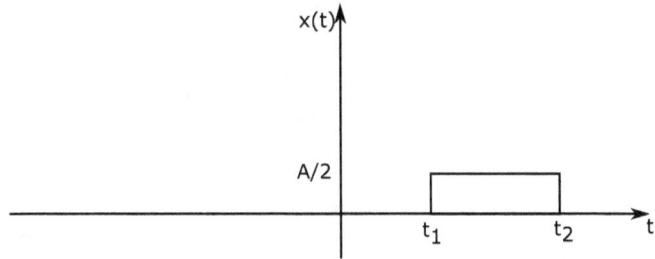

Figure 1.20: $w(t) = \frac{1}{2}x(t)$

1.15 Signal Operation - Time Shift

If $x(t)$ is a continuous time signal, then we can define a signal:

$$z(t) = x(t - t_0), \tag{1.46}$$

where t_0 is a constant real number. Note that:

- If $t_0 > 0$, i.e. a positive number, then $z(t)$ is shifted to the right compared to $x(t)$.

- If $t_0 < 0$, i.e. a negative number, then $z(t)$ is shifted to the left compared to $x(t)$.

Note that these operations are also defined for discrete time signals $x[n], z[n]$.

1.15.1 Solved Examples

1. Consider a signal given in Figure 1.18. Let $t_1 = 0, t_2 = 5$. Plot $z(t) = x(t - 3)$.

Solution: Comparing with Equation (1.46), $t_0 = 3$, therefore it is a shift to the right by 3 seconds. Therefore the plot of $z(t)$ is shown in Figure 1.21.

To verify the shift, we can check at a few time instances. This is shown in Table 1.1.

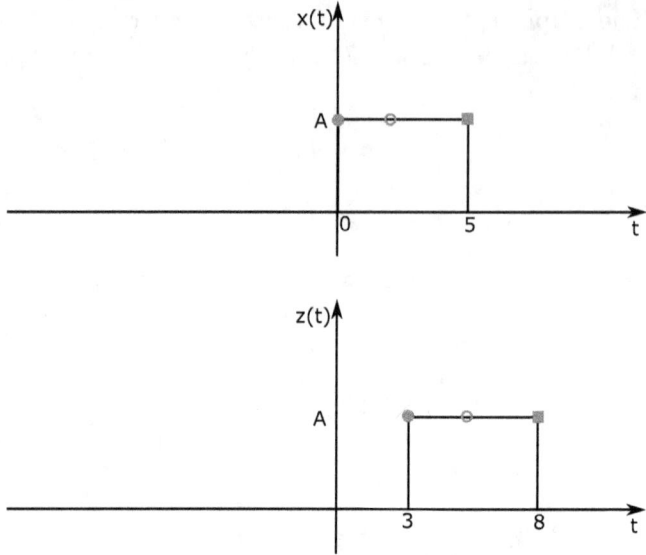

Figure 1.21: $z(t) = x(t - 3)$

t	$z(t)$
0	$z(0) = x(-3)$
3	$z(3) = x(0)$ (indicated by the red solid dot in Figure 1.21)
5	$z(5) = x(2)$ (indicated by the red circle in Figure 1.21)
8	$z(8) = x(5)$ (indicated by the red square in Figure 1.21)

Table 1.1: Verifying time shift to the right

2. Consider a signal given in Figure 1.18. Let $t_1 = 0, t_2 = 5$. Plot $z(t) = x(t + 3)$.

Solution: Comparing with Equation (1.46), $t_0 = -3$, therefore it is a shift to the left by 3 seconds. Therefore the plot of $z(t)$ is shown in Figure 1.22.

To verify the shift, we can check at a few time instances. This is shown in Table 1.2.

40

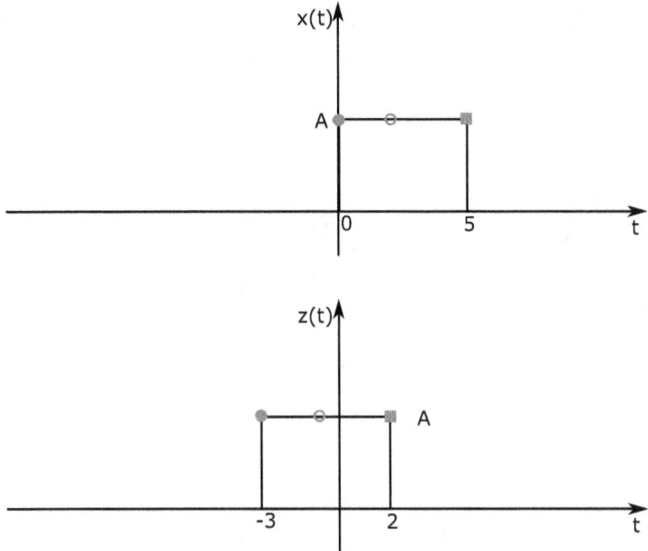

Figure 1.22: $z(t) = x(t+3)$

t	$z(t)$
-3	$z(-3) = x(0)$ (indicated by the red solid dot in Figure 1.22)
-1	$z(-1) = x(2)$ (indicated by the red circle in Figure 1.22)
2	$z(2) = x(5)$ (indicated by the red square in Figure 1.22)

Table 1.2: Verifying time shift to the left

1.16 Signal Operation - Time Reversal

If $x(t)$ is a continuous time signal, then we can define a signal:

$$z(t) = x(-t). \tag{1.47}$$

$z(t)$ is time-reversed version of $x(t)$. i.e. reflection about $t = 0$ (y-axis). Note that these operations are also defined for discrete time signals $x[n], z[n]$.

1.16.1 Solved Examples

1. Consider a signal given in Figure 1.18. Let $t_1 = 0, t_2 = 5$. Plot $p(t) = x(-t)$.

Solution: Comparing with Equation (1.47), $p(t)$ is a time reversed version of $x(t)$. Therefore the plot of $p(t)$ is shown in Figure 1.23.

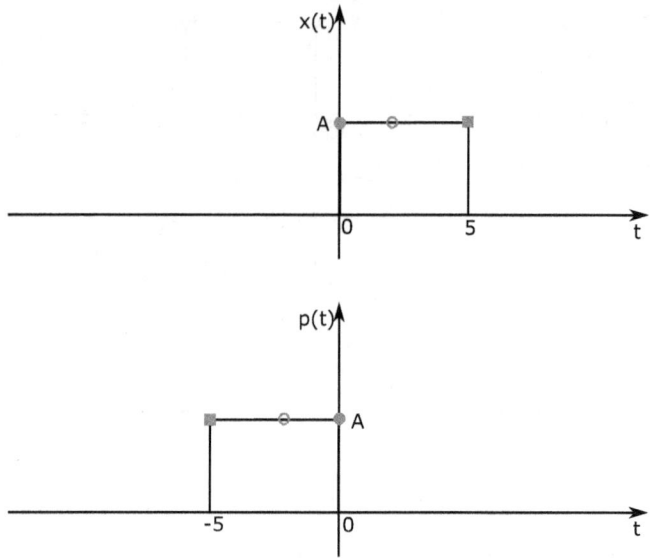

Figure 1.23: $p(t) = x(-t)$

To verify this, we can check at a few time instances. This is shown in Table 1.3.

t	$p(t)$
0	$p(0) = x(0)$ (indicated by the red solid dot in Figure 1.23)
-2	$p(-2) = x(2)$ (indicated by the red circle in Figure 1.23)
-5	$p(-5) = x(5)$ (indicated by the red square in Figure 1.23)

Table 1.3: Verifying time reversal

1.17 Signal Operation - Time Scale

If $x(t)$ is a CT signal, then we can define a signal:

$$z(t) = x(at), \qquad (1.48)$$

where a is a positive number. Note that':

- If $a > 1$, then $z(t)$ is a time-compressed version of $x(t)$.

42

- If $a < 1$, then $z(t)$ is a time-expanded version of $x(t)$.

Note that these operations are also defined for discrete time signals $x[n], z[n]$.

1.17.1 Solved Examples

1. Consider a signal given in Figure 1.18. Let $t_1 = 0, t_2 = T$. Plot $p(t) = x(2t)$.

Solution: $p(t)$ is a time-scaled version of $x(t)$ with $a > 1$. Therefore $p(t)$ is a time-compressed (by a factor of 2) version of $x(t)$. The plots of $x(t)$ and $p(t)$ are shown in Figure 1.24.

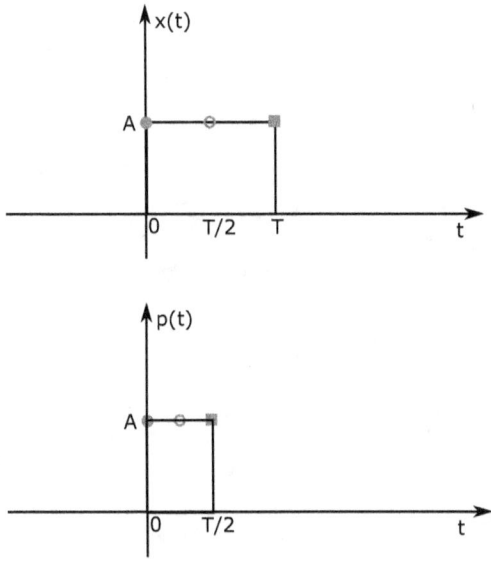

Figure 1.24: $p(t) = x(2t)$

To verify this, we can check at a few time instances. This is shown in Table 1.4.

2. Consider a signal given in Figure 1.18. Let $t_1 = 0, t_2 = T$. Plot $p(t) = x(\frac{1}{2}t)$.

Solution: $p(t)$ is a time-scaled version of $x(t)$ with $a < 1$. Therefore $p(t)$ is a time-expanded (by a factor of 2) version of $x(t)$. The plots of $x(t)$ and $p(t)$ are shown in Figure 1.25.

t	$p(t)$
0	$p(0) = x(0)$ (indicated by the red solid dot in Figure 1.24)
T/2	$p(T/2) = x(T)$ (indicated by the red square in Figure 1.24)
T/4	$p(T/4) = x(T/2)$ (indicated by the red circle in Figure 1.24)

Table 1.4: Verifying time scaling for $p(t) = x(2t)$.

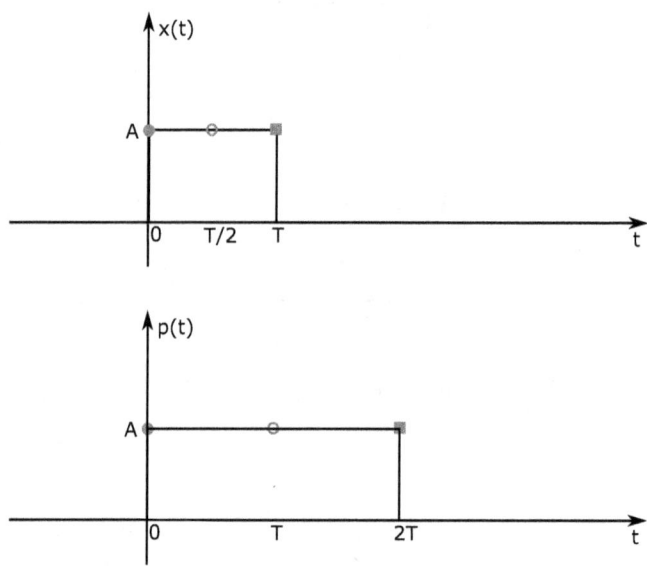

Figure 1.25: $p(t) = x(\frac{1}{2}t)$

To verify this, we can check at a few time instances. This is shown in Table 1.5.

t	$p(t)$
0	$p(0) = x(0)$ (indicated by the red solid dot in Figure 1.25)
T	$p(T) = x(T/2)$ (indicated by the red circle in Figure 1.25)
2T	$p(2T) = x(T)$ (indicated by the red square in Figure 1.25)

Table 1.5: Verifying time scaling for $p(t) = x(\frac{1}{2}t)$.

1.18 Combining Signal Operations

If $x(t)$ is a CT signal, then we can define a signal:

$$z(t) = x(\alpha t + \beta), \tag{1.49}$$

where α is a time-scaling factor, β is a time-shifting factor.

When plotting these signals, the rule is *first time-shift, then time-scale, then time-reveral*. Note that this applicable for discrete time signals also.

1.18.1 Solved Examples

1. Consider a signal given in Figure 1.26. Plot $p(t) = x(3 - 3t)$.

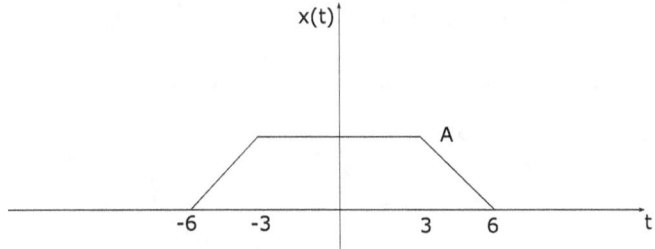

Figure 1.26: A signal $x(t)$

Solution: We can get $p(t)$ from $x(t)$ by combining the time shifting, time scaling and time reversal operations. To see this, define:

$$y(t) = x(t + 3)$$
$$y_1(t) = y(3t)$$
$$= x(3t + 3)$$
$$p(t) = y_1(-t)$$
$$= x(-3t + 3)$$

Thus to obtain the plot $p(t)$, we must plot $y(t)$ and $y_1(t)$. These are shown in Figure 1.27.

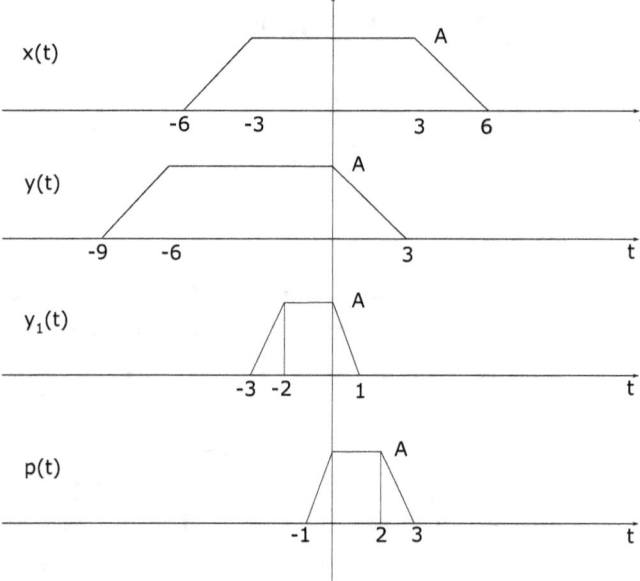

Figure 1.27: A signal $p(t) = x(3 - 3t)$

2. A signal $x(t)$ is given by:

$$x(t) = \begin{cases} 1, & -1 \leq t \leq 1 \\ 0, & \text{otherwise} \end{cases}$$

Plot $x(2(t - 2))$.

Solution: We can get the signal $p(t) = x(2(t - 2)) = x(2t - 4)$ by combining the time shifting and time scaling operations. To see this, define:

$$y_1(t) = x(t - 4)$$
$$p(t) = y_1(2t)$$
$$= x(2t - 4)$$

Thus to obtain the plot $p(t)$, we must plot $y_1(t)$. These are shown in Figure 1.28.

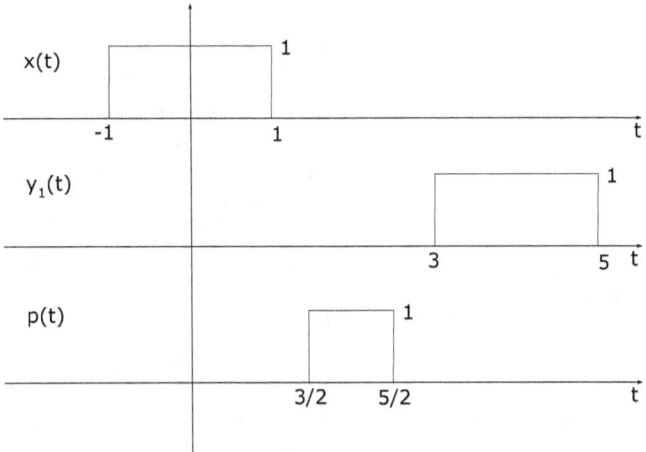

Figure 1.28: A signal $p(t) = x(2(t - 2))$

3. Plot the signal: $x(t) = u(t + 1) + 2u(t) - u(t - 3) - 2u(t - 5)$.

Solution: Using the time shifting and amplitude scaling operations and addition of signals, we get the plot of $x(t)$ shown in Figure 1.29.

1.19 Orthogonal and Orthonormal Signals

Two functions $u(t)$ and $v(t)$ are said to be *orthogonal* over the interval (a, b) if:

$$\int_a^b u(t)v^*(t)dt = 0. \tag{1.50}$$

If in addition to satisfying the condition in Equation (1.50), equations

$$\int_a^b |u(t)|^2 dt = 1 \tag{1.51}$$

47

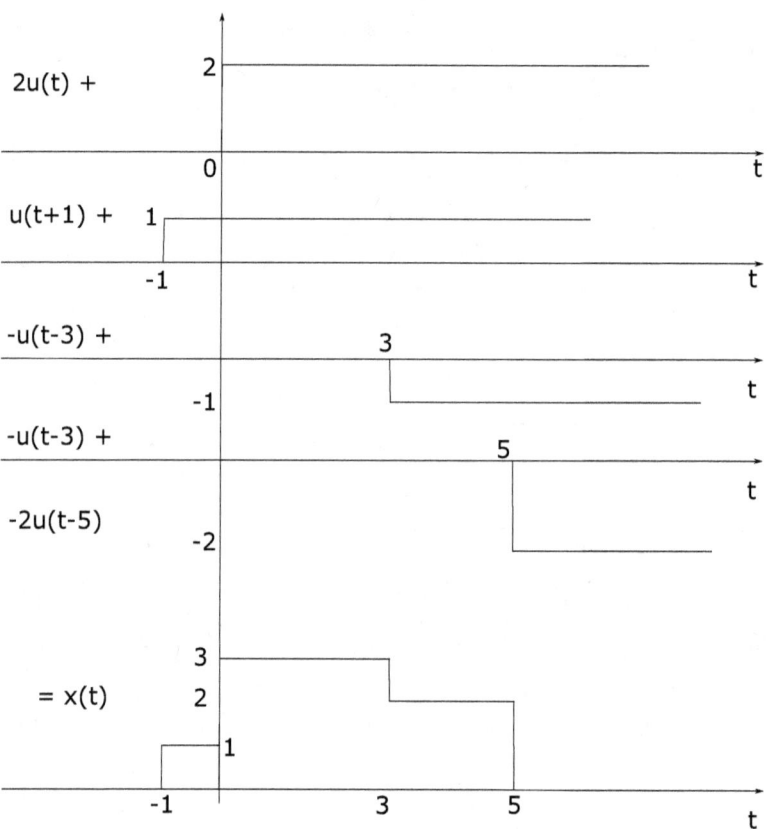

Figure 1.29: A signal $x(t)$

and

$$\int_a^b |v(t)|^2 dt = 1 \tag{1.52}$$

are satisfied, then the functions $u(t)$ and $v(t)$ are *orthonormal*.

Similarly for two discrete time signals $\phi_k[n]$ and $\phi_m[n](k \neq m)$, the orthogonality and orthonormality conditions are:

$$\sum_{n=N_1}^{N_2} \phi_k[n]\phi_m^*[n] = 0. \tag{1.53}$$

$$\sum_{n=N_1}^{N_2} |\phi_k[n]|^2 = \sum_{n=N_1}^{N_2} |\phi_m[n]|^2 = 1. \tag{1.54}$$

1.19.1 Solved Examples

1. Show that the two signals $u(t) = \sin(t)$, $v(t) = \cos(t)$ are orthogonal over the period $(0, 2\pi)$.

Solution: Applying Equation (1.50),

$$\int_0^{2\pi} \sin(t)\cos(t) dt = \int_0^{2\pi} \sin(2t)/2 \, dt$$

$$= \frac{1}{2}\int_0^{2\pi} \sin(2t) \, dt$$

$$= \frac{-1}{4}[\cos(2t)]_0^{2\pi}$$

$$= 0.$$

Therefore the signals are orthogonal.

2. Show that the two signals $\phi_k[n] = e^{jk\frac{2\pi}{N}n}$, $\phi_m[n] = e^{jm\frac{2\pi}{N}n}$ $(k \neq m)$ are orthogonal over an interval N.

Solution: Applying Equation (1.53),

$$\sum_{n=0}^{N-1} \phi_k[n]\phi_m^*[n] = \sum_{n=0}^{N-1} e^{jk\frac{2\pi}{N}n} e^{-jm\frac{2\pi}{N}n}$$

$$= e^{j(k-m)\frac{2\pi}{N}n}$$

Using the finite sum formula for the geometric series given by:

$$\sum_{n=0}^{N-1} \alpha^n = \frac{1-\alpha^N}{1-\alpha}, \alpha \neq 1 \qquad (1.55)$$

we get,

$$\sum_{n=0}^{N-1} \phi_k[n]\phi_m^*[n] = \frac{1 - e^{j\frac{2\pi}{N}(k-m)N}}{1 - e^{j\frac{2\pi}{N}(k-m)}}$$

$$= \frac{1 - e^{j2\pi(k-m)}}{1 - e^{j\frac{2\pi}{N}(k-m)}}$$

$$= 0. \ (\text{since } e^{j2\pi(k-m)} = 1)$$

Therefore the signals are orthogonal.

1.20 Correlation between Signals

If $x(t)$ and $y(t)$ are two real-valued signals, then the *cross-correlation function* is defined as:

$$\phi_{xy}(t) = \int_{-\infty}^{\infty} x(t+\tau)y(\tau)d\tau \qquad (1.56)$$

If $x(t)$ and $y(t)$ are two complex-valued signals, then the cross-correlation function is defined as:

$$\phi_{xy}(t) = \int_{-\infty}^{\infty} x(t+\tau)y^*(\tau)d\tau \qquad (1.57)$$

If $x(t)$ is a real-valued signal, then the *auto-correlation function* is defined as:

$$\phi_{xx}(t) = \int_{-\infty}^{\infty} x(t+\tau)x(\tau)d\tau \qquad (1.58)$$

2 Systems

Broadly, a system can be understood as a device (or a set of devices) that converts an input signal to an output signal. A system can also have multiple inputs and multiple outputs. A system can also be an interconnection of components or subsystems. Examples of systems include audio system, braking system, thermometer, etc. This chapter aims to answer the following questions:

1. How do we represent a system?

2. What are the various types of systems?

3. What is meant by the impulse response of a system?

4. How are the input and output signals in a LTI system related?

5. What is the concept of convolution between signals?

6. How do we classfiy a LTI systems using the impulse response of the system?

The concept map for the topics covered in this chapter is shown in Figure 2.1.

2.1 Representing Systems

One way to represent a continuous time system is shown in Figure 2.2. Similarly a discrete time system is represented as shown in Figure 2.3.

This is also represented as

$$x(t) \xrightarrow{S} y(t)$$
$$x[n] \xrightarrow{S} y[n],$$

where S represents the system, $x(t)$ and $x[n]$ are the input to the system and $y(t)$ and $y[n]$ are the output of the system.

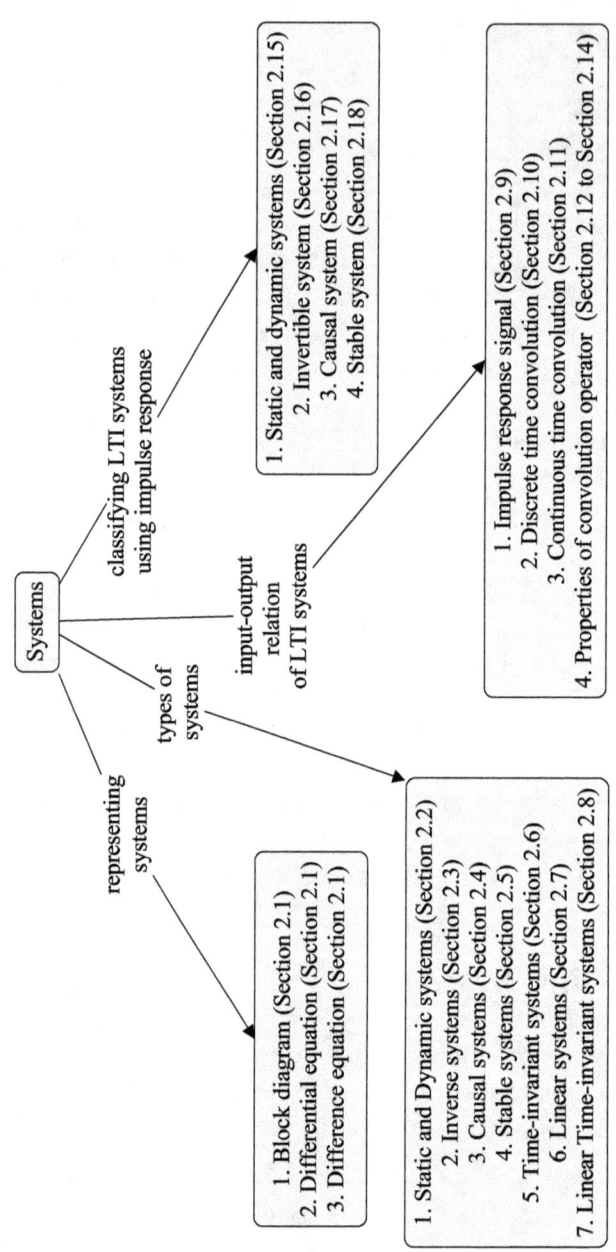

Figure 2.1: Concept map of topics in Chapter 2

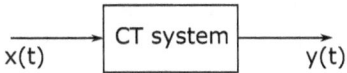

Figure 2.2: Continuous time system

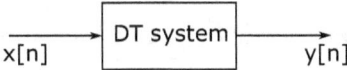

Figure 2.3: Discrete time system

Another way to represent a continuous time system is using a differential equation. Examples of such equations are shown below.

$$y(t) = x(t)$$

$$\frac{d}{dt}y(t) + ay(t) = x(t)$$

$$\frac{d^2}{dt^2}y(t) + 4\frac{d}{dt}y(t) + 3y(t) = \frac{d}{dt}x(t) + 2x(t)$$

Similarly, a discrete time system is represented using a difference equation. Examples of such equations are shown below.

$$y[n] = x[n]$$

$$y[n] - ay[n-1] = x[n]$$

$$y[n] - 3/4y[n-1] + 1/8y[n-2] = 2x[n]$$

Both differential and difference equations give a mathematical representation of how the system operates. That is, it gives a relation linking the input ($x(t)$ or $x[n]$) and the output ($y(t)$ or $y[n]$) of the system.

2.2 Static and Dynamic Systems

2.2.1 Static (memoryless) Systems

Definition: A system is said to be memoryless if its output at a given time is dependent only on the input at the given time. i.e. output is determined entirely by the present input only.

For example, consider a resistor R (shown in Figure 2.4) with the current $i(t)$ as the input and the voltage drop across it $v(t)$ as the output. The input-output relation is given by $v(t) = Ri(t)$. This is an example of a static system since $v(0)$

Figure 2.4: Example of a static system - resistor

depends only on $i(0)$, $v(2.5)$ depends only on $i(2.5)$ and so on; that is the output at a certain time depends only on the input at that time instant.

Other examples of static systems are: $y[n] = x[n]$, $y[n] = (2x[n] - x^2[n])^2$.

2.2.2 Dynamic (with memory) Systems

Definition: if the system output depends on the past or future values of the input, then the system is a dynamic system or system with memory.

For example, consider a delay system shown in Figure 2.5. The system is represented by the following input-output relation: $y[n] = x[n - 3]$. This is an example of a dynamic system since $y[3]$ depends on $x[0]$, $y[8]$ depends on $x[5]$ and so on; that is the output at a certain time depends on the input at a past time instant.

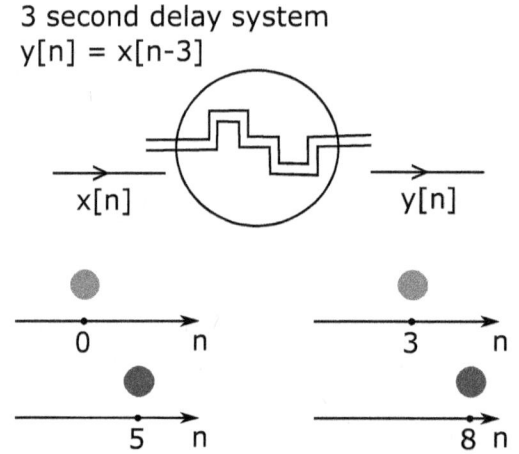

Figure 2.5: Example of a dynamic system - delay

Other examples of dynamic systems include: $y[n] = x[n-1]$, $y[n] = y[n-1]+x[n]$, $y(t) = \frac{1}{C}\int_{-\infty}^{t} x(\tau)d\tau$, $y[n] = \sum_{k=-\infty}^{n} x[k]$.

2.2.3 Solved Examples

1. Is the system $y(t) = x(2t)$, a static or dynamic system?

Solution: Checking the output at a few instances (Table 2.1):

t	$y(t)$
0	$y(0) = x(0)$
2	$y(2) = x(4)$ (output depends on a future value)
-1	$y(-1) = x(-2)$ (output depends on a past value)

Table 2.1: Verifying if the system $y(t) = x(2t)$ is static or dynamic

Since the output depends on the past and future values, the system is a dynamic system (system with memory).

2. A discrete-time system is described by the following input-output relation: $y[n] = nx[n]$. Is this system static or dynamic?

Solution: Checking the output at a few instances (Table 2.2):

n	$y[n]$
0	$y[0] = 0$
1	$y[1] = x[1]$
2	$y[2] = 2x[2]$
-3	$y[-3] = (-3)x[-3]$

Table 2.2: Verifying if the system $y[n] = nx[n]$ is static or dynamic

Since the output depends only on the present value of the input, the system is a static system.

2.3 Inverse Systems

Definition: A system is said to be invertible if distinct inputs lead to distinct outputs. If a system is invertible, then an inverse system exists, such that, when cascaded with the original system, yields an output $w[n]$ equal to the input $x[n]$ to the first system. This is shown in Figure 2.6.

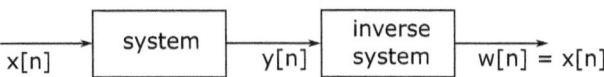

Figure 2.6: Inverse system

2.3.1 Solved Examples

1. Is the system $y(t) = 2x(t)$ invertible?

Solution: If we define an inverse system with the following input-output relation $w(t) = 1/2\ y(t)$; then we get:

$$w(t) = \frac{1}{2}(2x(t))$$
$$= x(t).$$

Therefore since an inverse system exists the original system is said to be invertible.

2. Is the following system invertible?

$$y[n] = \sum_{k=-\infty}^{n} x[k]$$

Solution: $y[n]$ is given by:

$$y[n] = x[-\infty] + \cdots + x[n-2] + x[n-1] + x[n]$$

Similarly $y[n-1]$ is given by:

$$y[n-1] = x[-\infty] + \cdots + x[n-2] + x[n-1]$$

If we define a an inverse system with the following input-output relation $w[n] = y[n] - y[n-1]$, then we get $w[n] = x[n]$. Therefore since an inverse system exists the original system is said to be invertible.

3. Is the system $y[n] = 0$ invertible?

Solution: This system does not satisfy the condition that distinct input leads to distinct output. For example, regardless of the input signal $x_1[n]$ or $x_2[n]$, the output is always 0. Therefore this system is not invertible.

2.4 Causal Systems

Definition: A system is causal if the output at any time depends only on the values of the input at the present time and in the past.

Examples of causal systems include: $y[n] = \sum_{k=-\infty}^{n} x[k]$, $y(t) = \frac{1}{C}\int_{-\infty}^{t} x(\tau)d\tau$. Examples of non-causal systems: $y[n] = x[n] - x[n+1]$, $y(t) = x(t+1)$.

2.5 Stable Systems

In this book, when we discuss stability, we refer to bounded input bounded output (BIBO) stability. A bounded signal is a signal whose magnitude does not grow without bound i.e. there is an upper limit.

Figure 2.7 shows an example of a bounded signal, $y(t) = 2$. In this case we are able to define an upper bound for the signal, e.g. $y(t)$ is below an upper bound 5.

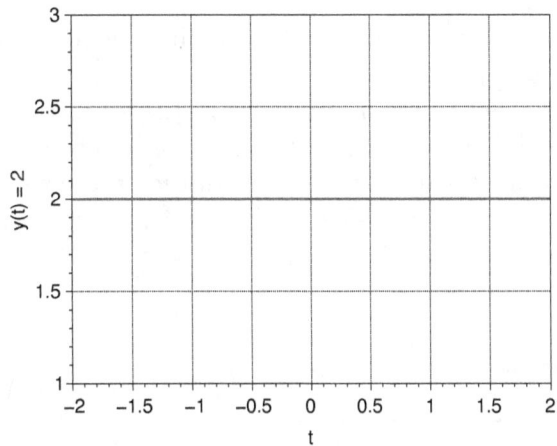

Figure 2.7: A bounded signal, $y(t) = 2$

Figure 2.8 shows an example of an unbounded signal, $y(t) = e^{2t}$. In this case we are unable to define an upper bound for the signal. For example, if we say that $y(t)$ has an upper bound 20, then this is not true since at a time $t = 2$, the value of the function is greater than this upper bound.

Definition: A BIBO stable system is a system where if the input is bounded then the output is also bounded (i.e. the output cannot diverge). An example of a stable system is a pendulum. If the push we give to the pendulum is considered as the input to the system and the angle which the pendulum makes from its resting position is considered as the output of the system, then this system is a stable system, as

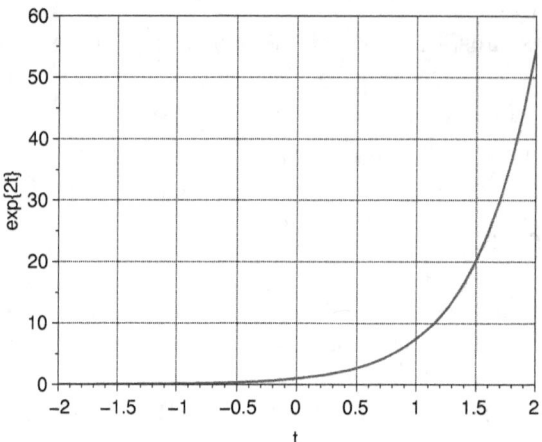

Figure 2.8: An unbounded signal, $y(t) = e^{2t}$

the output is always bounded (meaning, the pendulum swings to a finite angle and always comes to rest).

To solve problems related to stability, typcially there are two approaches. The first approach involves finding a counter example; meaning if we take a bounded input but the output of the system is not bounded, then the system is said to be not stable. The second approach is more of a mathematical approach where we consider all bounded inputs and show that the output is also bouned. Both these approaches are seen in the examples below.

2.5.1 Solved Examples

1. Is the system $y(t) = tx(t)$ stable?

Solution: For this problem we will utilize the first approach. Consider an input $x(t) = 1$, which is a bounded signal. The output of the system becomes $y(t) = t$. This is shown in Figure 2.9.

This output signal $y(t) = t$ is an unbounded signal, since an upper bound cannot be defined. Therefore this system is unstable (because for a bounded input the system gives an unbounded ouput).

2. Is the system $y(t) = e^{x(t)}$ stable?

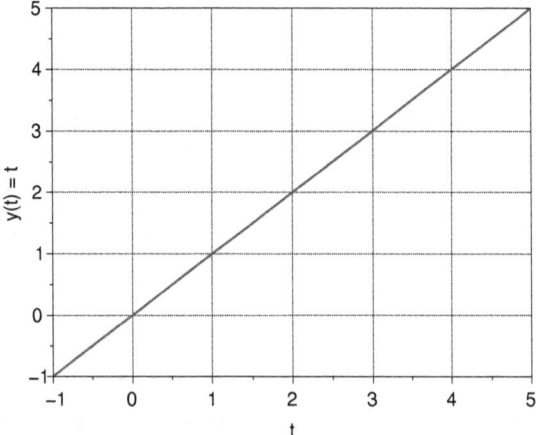

Figure 2.9: An unbounded signal, $y(t) = t$

Solution: For this problem we will utilize the second approach, since it is not easy to find a counter example. Assume that the input signal is bounded with an upper bound B. That is

$$|x(t)| < B \; \forall t$$
$$-B < x(t) < B$$
$$e^{-B} < e^{x(t)} < e^{B}$$
$$e^{-B} < y(t) < e^{B}$$
$$|y(t)| < e^{B} \; \forall t$$

That is, the output $y(t)$ is bounded for all possible bounded input $x(t)$. Therefore the system is stable.

3. Test the stability of the following system: $y(t) = x(-t - 2)$.

Solution: For this problem we will utilize the second approach. First consider a system $y(t) = x(t)$. Assume that the input signal is bounded with an upper bound B. That is

$$|x(t)| < B \; \forall t$$
$$|x(t-2)| < B \; \forall t \quad \text{(time-shifted input signal is also bounded)}$$
$$|x(-t-2)| < B \; \forall t \quad \text{(time-reversed signal is also bounded)}$$
$$|y(t)| < B \; \forall t$$

That is, the output $y(t)$ is bounded for all possible bounded input $x(t)$. Therefore the system is stable.

2.6 Time-invariant Systems

Definition: A system is time-invariant if a time-shift in the input signal results in an identical time-shift in the output signal.

For example, consider the delay system seen earlier in Figure 2.5. The system is represented by the following input-output relation: $y[n] = x[n-3]$. This is an example of a time-invariant system since when the input was delayed by 5 seconds ($x[5]$), the output was also delayed by 5 seconds ($y[8]$).

To solve problems related to time-invariance, we will be using a two-path approach. If the output from the two paths are the same, the system is said to be time-invariant.

The first path is shown in Figure 2.10. Here the input signal $x(t)$ is the input to the system and the output signal is $y(t)$. Then $y(t)$ goes through a delay system and the output is given by $y_d(t) = y(t - t_0)$.

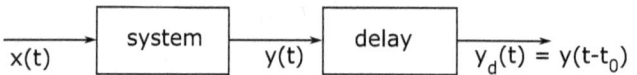

Figure 2.10: Checking time-invariance - path 1

In the second path (shown in Figure 2.11), the input signal $x(t)$ is first sent through a delay system. The output is $x_d(t)$. This signal is then input to the system, which gives an output $y'(t)$.

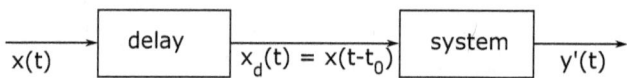

Figure 2.11: Checking time-invariance - path 2

If $y_d(t) = y'(t)$, then the system is said to be time-invariant (meaning, when the input is delayed by a certain amount, the output is also delayed by the same amount). If not, then the system is time-varying or not time-invariant. Note that in both paths, the delay system delays the signal by the same amount t_0.

2.6.1 Solved Examples

1. Is the system $y(t) = x(2t)$, a time-varying or time-invariant system?

Solution: Following the two-path approach discussed above, from path 1 we get:

$$x(t) \xrightarrow{S} y(t) = x(2t)$$
$$\downarrow \text{delay}$$
$$y_d(t) = y(t - t_0) = x(2(t - t_0)) = x(2t - 2t_0)$$

From path 2 we get:

$$x(t) \xrightarrow{\text{delay}} x_d(t) = x(t - t_0)$$
$$\downarrow S$$
$$y'(t) = x_d(2t) = x(2t - t_0)$$

Since $y_d(t) \neq y'(t)$, the system is not time-invariant.

2. A discrete-time system is described by the following input-output relation: $y[n] = nx[n]$. Is this system time-varying or time-invariant?

Solution: Following the two-path approach discussed above, from path 1 we get:

$$x[n] \xrightarrow{S} y[n] = nx[n]$$
$$\downarrow \text{delay}$$
$$y_d[n] = y[n - n_0] = (n - n_0)x[n - n_0]$$

From path 2 we get:

$$x[n] \xrightarrow{\text{delay}} x_d[n] = x[n - n_0]$$
$$\downarrow S$$
$$y'[n] = nx_d[n] = nx[n - n_0]$$

Since $y_d[n] \neq y'[n]$, the system is not time-invariant.

2.7 Linear Systems

Definition: A linear system is a system that satisfies the property of superposition. The superposition property consists of two parts: (a) homogenity and (b) additivity.

A system S is said to satisfy the homogenity property, if:

$$x(t) \xrightarrow{S} y(t) \text{ then,}$$

$$\alpha x(t) \xrightarrow{S} \alpha y(t)$$

where α is a constant. That is, if the input is scaled by a constant, then the ouput signal is also scaled by the same factor.

A system S is said to satisfy the additivity property, if:

$$x_1(t) \xrightarrow{S} y_1(t) \text{ and,}$$

$$x_2(t) \xrightarrow{S} y_2(t) \text{ then,}$$

$$x_1(t) + x_2(t) \xrightarrow{S} y_1(t) + y_2(t).$$

Thus if a system has to satisfy the superposition property, it must satisfy both the homogenity and additivity properties. Therefore if a system S is said to satisfy the superposition property if:

$$x_1(t) \xrightarrow{S} y_1(t) \text{ and,}$$

$$x_2(t) \xrightarrow{S} y_2(t) \text{ then,}$$

$$\alpha x_1(t) + \beta x_2(t) \xrightarrow{S} \alpha y_1(t) + \beta y_2(t)$$

where α and β are constants.

2.7.1 Solved Examples

1. Is the system $y(t) = x(2t)$, a linear or non-linear system?

Solution: Consider two inputs $x_1(t)$ and $x_2(t)$ such that:

$$x_1(t) \xrightarrow{S} y_1(t) = x_1(2t) \text{ and,}$$

$$x_2(t) \xrightarrow{S} y_2(t) = x_2(2t).$$

Let us check what is the output of the system S for an input $x_3(t) = \alpha x_1(t) + \beta x_2(t)$.

$$x_3(t) = \alpha x_1(t) + \beta x_2(t) \xrightarrow{S} y_3(t) = x_3(2t)$$
$$y_3(t) = \alpha x_1(2t) + \beta x_2(2t)$$
$$x_3(t) \xrightarrow{S} y_3(t) = \alpha y_1(t) + \beta y_2(t)$$

Therefore the system S satisfies the superposition property. Therefore the system is linear.

> 2. A system has the input-output relation given by: $y = x^2$. Show that this system is non-linear.

Solution: Consider two inputs $x_1(t)$ and $x_2(t)$ such that:

$$x_1(t) \xrightarrow{S} y_1(t) = x_1^2(t)$$
$$x_2(t) \xrightarrow{S} y_2(t) = x_2^2(t).$$

Let us check what is the output of the system S for an input $x_3(t) = \alpha x_1(t) + \beta x_2(t)$.

$$x_3(t) = \alpha x_1(t) + \beta x_2(t) \xrightarrow{S} y_3(t) = x_3^2(t)$$
$$y_3(t) = [\alpha x_1(t) + \beta x_2(t)]^2$$
$$y_3(t) = \alpha^2 x_1^2(t) + \beta^2 x_2^2(t) + 2\alpha\beta x_1(t)x_2(t)$$
$$y_3(t) = \alpha^2 y_1(t) + \beta^2 y_2(t) + 2\alpha\beta x_1(t)x_2(t)$$
$$x_3(t) \xrightarrow{S} y_3(t) \neq \alpha y_1(t) + \beta y_2(t)$$

Therefore the system is non-linear.

> 3. A discrete-time system is described by the following input-output relation: $y[n] = nx[n]$. Is this system linear or non-linear?

Solution: Consider two inputs $x_1[n]$ and $x_2[n]$ such that:

$$x_1[n] \xrightarrow{S} y_1[n] = nx_1[n]$$
$$x_2[n] \xrightarrow{S} y_2[n] = nx_2[n].$$

Let us check what is the output of the system S for an input $x_3[n] = \alpha x_1[n] + \beta x_2[n]$.

$$x_3[n] = \alpha x_1[n] + \beta x_2[n] \xrightarrow{S} y_3[n] = nx_3[n]$$
$$y_3[n] = n[\alpha x_1[n] + \beta x_2[n]]$$
$$y_3[n] = n\alpha x_1[n] + n\beta x_2[n]$$
$$x_3[n] \xrightarrow{S} y_3[n] = \alpha y_1[n] + \beta y_2[n].$$

Therefore the system S satisfies the superposition property. Therefore the system is linear.

2.8 Linear Time-invariant (LTI) Systems

This book will be focusing on LTI systems, as many physical (or natural) systems can be modelled this way. LTI systems are linear systems (Section 2.7) as well as time-invariant systems (Section 2.6).

2.9 Impulse Response Signal

In Section 1.2 we have the seen the unit impulse signal ($\delta(t)$) or the unit sample signal ($\delta[n]$). The response of a system to an impulse function ($\delta(t)$ or $\delta[n]$) is called the impulse response signal (or function). It is denoted by $h(t)$ for CT systems and $h[n]$ for DT systems. This is shown in Figure 2.12.

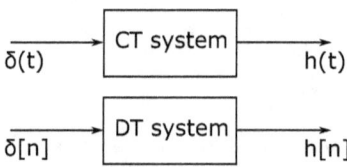

Figure 2.12: Impulse response function

Now if the system is time-invariant, then if a time-shifted impulse function is the input to a time-invariant system then the output of the system is a time-shifted impulse response. Note that both the input and the output signals are shifted by the same amount k. This is shown in Figure 2.13.

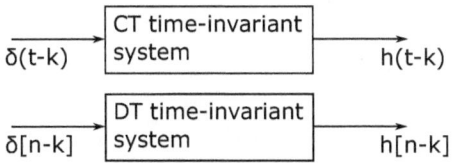

Figure 2.13: Time-shifted impulse response

2.9.1 Solved Example

1. A discrete-time LTI system model of a two-path propagation channel is given by:
$$y[n] = x[n] + \frac{1}{2}x[n-1].$$

   What is the impulse response of the system?

Solution: By definition, when $x[n] = \delta[n]$, $y[n] = h[n]$. Therefore the impulse response is given by $h[n] = \delta[n] + \frac{1}{2}\delta[n-1]$.

2.10 Input-Output relation for DT LTI Systems (Discrete Time Convolution)

Now that we have seen an LTI system, we are interested to answer the following question: if there is an input signal $x[n]$ to a discrete time LTI system, what will the output signal $y[n]$ look like?

To anwer this question or to derive an expression for $y[n]$, a derivation is required.

1. The input $x[n]$ is represented as a linear combination of shifted impulse functions ($\delta[n]$) as shown:
$$x[n] = \sum_{k=-\infty}^{\infty} x[k]\delta[n-k] \tag{2.1}$$

In order to understand why this is true, see the example shown in Figure 2.14. Here, the signal $x[n]$ is written as a linear combination of three time-shifted impulse functions.

Expanding Equation (2.1), we can write:
$$\begin{aligned} x[n] &= \ldots + x[-100]\delta[n+100] + \ldots + x[-1]\delta[n+1] + \\ &\quad x[0]\delta[n] + \\ &\quad x[1]\delta[n-1] + \ldots + x[100]\delta[n-100] + \ldots \end{aligned} \tag{2.2}$$

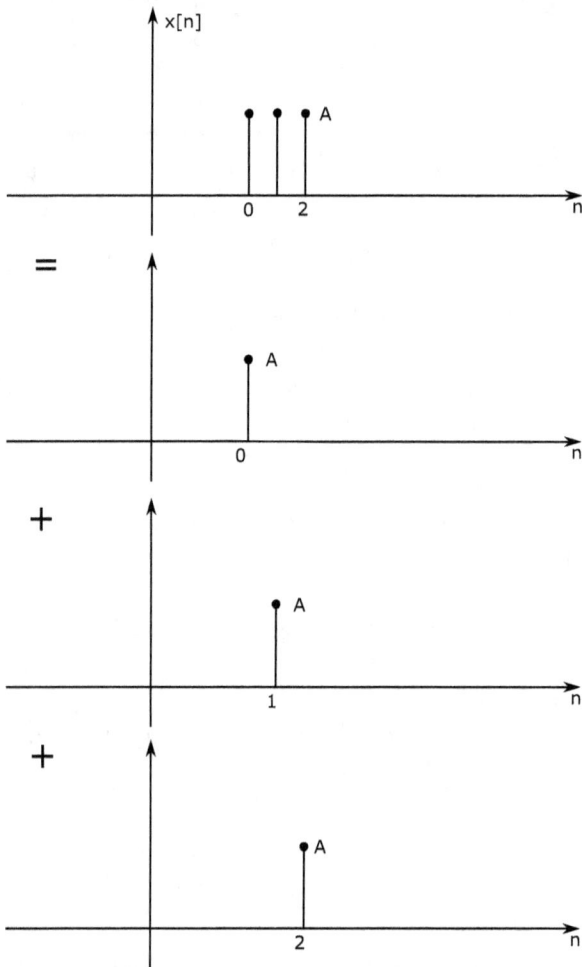

Figure 2.14: e.g. x[n] represented as a sum of time-shifted impulse functions

2. Since the system is time-invariant (see Figure 2.13), the following is true:

$$\delta[n - k] \longrightarrow h[n - k]. \tag{2.3}$$

3. Also, since the system is linear, the following are true:

$$x[-100]\delta[n + 100] \longrightarrow x[-100]h[n + 100] \text{ (homogenity property)} \tag{2.4}$$

$$\delta[n + 1] + \delta[0] \longrightarrow h[n + 1] + h[0] \text{ (additivity property)} \tag{2.5}$$

4. Applying Equations (2.3), (2.4), (2.5) to Equation (2.2), the output $y[n]$ from the DT LTI system is given by:

$$
\begin{aligned}
y[n] \quad &= \ldots + x[-100]h[n + 100] + \ldots + x[-1]h[n + 1] + \\
&\quad x[0]h[n] + \\
&x[1]h[n - 1] + \ldots + x[100]h[n - 100] + \ldots.
\end{aligned} \tag{2.6}
$$

That is,

$$y[n] = \sum_{k=-\infty}^{\infty} x[k]h[n - k], \tag{2.7}$$

meaning, the output $y[n]$ is represented as a linear combination of *impulse response* functions.

5. Equation (2.7) is called the convolution sum and can be written as:

$$y[n] = x[n] * h[n], \tag{2.8}$$

where $*$ is used to denote the convolution operation.

Note that the impulse response signals do not depend on the input to the system. Thus Equation (2.7) tells us that if we know the impulse response $h[n]$ then for any input signal $x[n]$, we can compute the output $y[n]$ using the convolution operator.

2.11 Input-Output relation for CT LTI Systems (Continuous Time Convolution)

The previous section discussed about discrete time LTI systems. In this section, a continuous time LTI system is considered. We are interested to answer the following question: if there is an input signal $x(t)$ to a CT LTI system, what will the output signal $y(t)$ look like? To derive this, we will take an approach similar to that discussed in the previous section.

1. Similar to Equation (2.1), for CT systems the input $x(t)$ is represented as a linear combination of shifted impulse functions ($\delta(t)$) as shown:

$$x(t) = \int_{-\infty}^{\infty} x(\tau)\delta(t - \tau)d\tau \qquad (2.9)$$

2. Since the system is linear and time-invariant, Equations (2.3), (2.7), (2.8) in discrete time, become:

$$\delta(t - \tau) \longrightarrow h(t - \tau). \qquad (2.10)$$

$$y(t) = \int_{-\infty}^{\infty} x(\tau)h(t - \tau)d\tau. \qquad (2.11)$$

$$y(t) = x(t) * h(t), \qquad (2.12)$$

where $h(t)$ is the impulse response function. Note again that the impulse response signals do not depend on the input to the system. Thus Equation (2.12) tells us that if we know the impulse response $h(t)$ then for any input signal $x(t)$, we can compute the output $y(t)$ using the convolution operator.

2.12 Commutative Property of the Convolution Operator

The commutative property of the convolution operation in the discrete-time and continuous-time domains are given by:

$$x[n] * h[n] = h[n] * x[n] \qquad (2.13)$$

$$x(t) * h(t) = h(t) * x(t) \qquad (2.14)$$

A block diagram approach of understanding the commutative property (assuming a discrete-time system) is shown in Figure 2.15. The top block diagram shows the left-hand side of the above equation, while the bottom block diagram shows the right-hand side of the above equation.

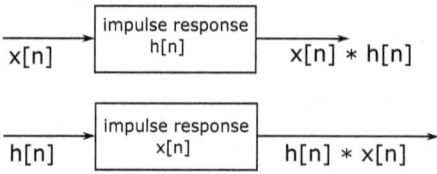

Figure 2.15: Commutative property of the convolution operator

2.13 Distributive Property of the Convolution Operator

The distributive property of the convolution operation in the discrete-time and continuous-time domains are given by:

$$x[n] * (h_1[n] + h_2[n]) = x[n] * h_1[n] + x[n] * h_2[n] \qquad (2.15)$$

$$x(t) * (h_1(t) + h_2(t)) = x(t) * h_1(t) + x(t) * h_2(t) \qquad (2.16)$$

A block diagram approach of understanding the distributive property (assuming a discrete-time system) is shown in Figure 2.16. The top block diagram shows the left-hand side of the above equation, while the bottom block diagram shows the right-hand side of the above equation.

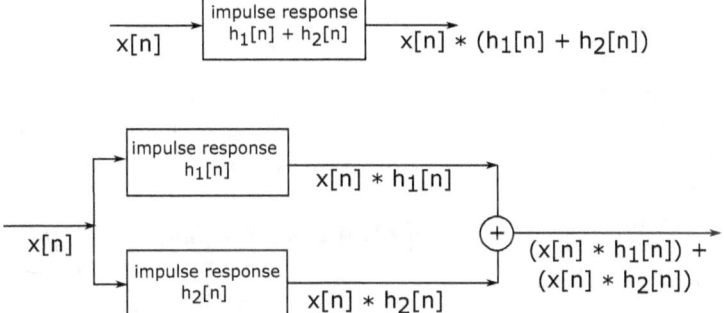

Figure 2.16: Distributive property of the convolution operator

2.14 Associative Property of the Convolution Operator

The associate property of the convolution operation in the discrete-time and continuous-time domains are given by:

$$x[n] * (h_1[n] * h_2[n]) = (x[n] * h_1[n]) * h_2[n] \qquad (2.17)$$

$$x(t) * (h_1(t) * h_2(t)) = (x(t) * h_1(t)) * h_2(t) \qquad (2.18)$$

A block diagram approach of understanding the associative property (assuming a discrete-time system) is shown in Figure 2.17. The top block diagram shows the left-hand side of the above equation, while the bottom block diagram shows the right-hand side of the above equation.

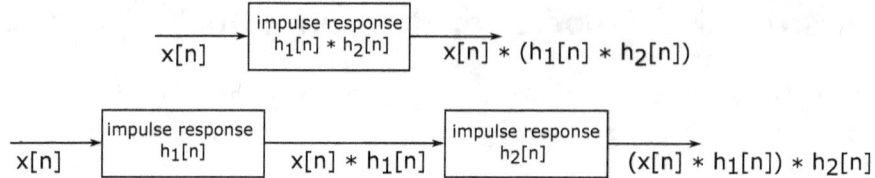

Figure 2.17: Associate property of the convolution operator

2.15 Static and Dynamic Property of LTI Systems

In this section, we are interested to answer the following question: if a LTI system is to be static or dynamic (refer Section 2.2), what condition should the impulse response ($h[n]$ or $h(t)$) satisfy?

Derivation: For a DT LTI system, with input $x[n]$ and impulse response $h[n]$, the output $y[n]$ is given by

$$y[n] = x[n] * h[n]$$

$$= \sum_{k=-\infty}^{\infty} x[k]h[n-k]$$

$$= \ldots + x[-2]h[n+2] + x[-1]h[n+1] + x[0]h[n] + x[1]h[n-1] + x[2]h[n-2] + \ldots$$

For a system without memory, $y[n]$ depends only on $x[n]$.

- For $n = -1$, y[-1] should depend only on $x[-1]$. That is, from the above equation, we want $h[n+1] \neq 0$ and all other terms $h[n+2] = h[n] = h[n-1] = \ldots = 0$. For $n = -1, h[n+1] = h[-1+1] = h[0] \neq 0$.

- For $n = 0$, y[0] should depend only on $x[0]$. That is, from the above equation, we want $h[n] \neq 0$ and all other terms $h[n+1] = h[n-1] = \ldots = 0$. For $n = 0, h[n] = h[0] \neq 0$.

- For $n = 1$, y[1] should depend only on $x[1]$. That is, from the above equation, we want $h[n-1] \neq 0$ and all other terms $h[n+1] = h[n] = h[n-2] = \ldots = 0$. For $n = 1, h[n-1] = h[1+1] = h[0] \neq 0$.

Generalizing the above the condition on $h[n]$ is given by:

$$h[n] = \begin{cases} 0, & n \neq 0 \\ 0, & n = 0 \end{cases}$$

Equivalently, $h[n]$ can be written as $h[n] = h[0]\delta[n]$.

To summarize,

- For a DT system without memory,

$$h[n] = h[0]\delta[n] \tag{2.19}$$

- For a CT system without memory,

$$h(t) = h(0)\delta(t) \tag{2.20}$$

- If the LTI system has an impulse response that is not zero for $n \neq 0$ or $t \neq 0$, then the system has memory.

2.16 Invertibility of LTI Systems

In this section, we are interested to answer the following question: if a LTI system is to be causal (refer Section 2.3), what condition should the impulse response ($h[n]$ or $h(t)$) satisfy?

Consider a discrete-time system with an impulse response $h[n]$ with input $x[n]$ and output $y[n]$. Let another LTI system with an impulse response $h_1[n]$ be cascaded with the first system as shown in Figure 2.18. The input to the second system is $y[n]$ and the output is $w[n]$.

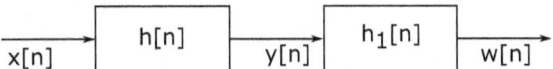

Figure 2.18: A discrete-time system

The output $w[n]$ is given by

$$w[n] = y[n] * h_1[n]$$
$$= x[n] * h[n] * h_1[n]. \text{ (since } y[n] = x[n] * h[n])$$

If we want the second system to be the inverse of the first system, then we want $w[n] = x[n]$. This is possible only if $h[n] * h_1[n] = \delta[n]$ (since $w[n] = x[n] * \delta[n] = x[n]$).

To summarize,

- For a DT LTI system with impulse response $h[n]$ is to be invertible, then there must exist a system with impulse response $h_1[n]$ such that the following equation is satisfied:

$$h[n] * h_1[n] = \delta[n] \tag{2.21}$$

- Similarly, for a CT LTI system with impulse response $h(t)$ is to be invertible, then there must exist a system with impulse response $h_1(t)$ such that the following equation is satisfied:

$$h(t) * h_1(t) = \delta(t) \tag{2.22}$$

2.17 Causality of LTI Systems

In this section, we are interested to answer the following question: if a LTI system is to be causal (refer Section 2.5), what condition should the impulse response ($h[n]$ or $h(t)$) satisfy?

Derivation: For a DT LTI system, with input $x[n]$ and impulse response $h[n]$, the output $y[n]$ is given by

$$y[n] = x[n] * h[n]$$
$$= \sum_{k=-\infty}^{\infty} x[k]h[n-k]$$
$$= \ldots + x[-2]h[n+2] + x[-1]h[n+1] + x[0]h[n] + x[1]h[n-1] +$$
$$x[2]h[n-2] + \ldots$$

From the above equation, for a causal system:

- For $n = 0$, $y[0]$ is to depend on $x[0], x[-1], x[-2], \ldots$. That is, we want $h[n-1] = h[n-2] = h[n-3] = \ldots = 0$. That is, $h[n-k] = 0$ for $k > 0$.

- For $n = 1$, $y[1]$ is to depend on $x[1], x[0], x[-1], x[-2], \ldots$. That is, we want $h[n-2] = h[n-3] = \ldots = 0$. That is, $h[n-k] = 0$ for $k > 1$.

Thus, in general, we want $h[n-k] = 0$ for $k > n$ (or $n - k < 0$).

Using a change of variables, if $n - k = m$, then the condition for a DT LTI system to be causal is given by $h[m] = 0$ for $m < 0$.

To summarize,

- For a DT LTI system with impulse response $h[n]$ to be causal, it must satisfy the following condition:

$$h[n] = 0, \text{for } n < 0 \tag{2.23}$$

- Similarly, for a CT LTI system with impulse response $h(t)$ to be causal, it must satisfy the following condition:

$$h(t) = 0, \text{for } t < 0 \tag{2.24}$$

2.17.1 Solved Examples

> 1. A LTI system has an impulse response given by: $h[n] = \alpha^n u[n]$. Is the system causal?

Solution: We would like to check if $h[n] = 0$, for $n < 0$.

From the definition of $u[n]$, $u[n] = 0, n < 0$. Thus $h[n] = 0, n < 0$. Thus the system is causal.

> 2. Check the causality of the LTI system with impulse response $h(t) = e^{-2t}u(t+2)$.

Solution: The plot of $h(t)$ is shown in Figure 2.19. For causality, $h(t) = 0$, for $t < 0$. This is not true in this case. Therefore the system is not causal.

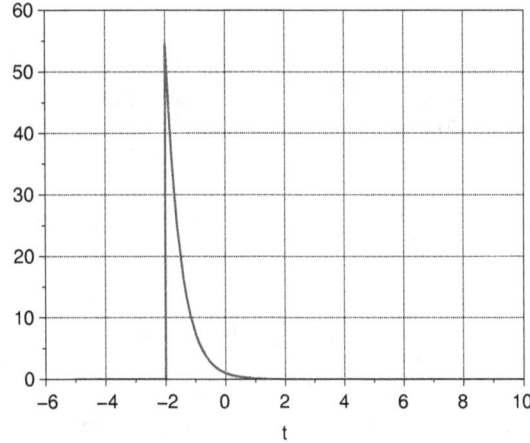

Figure 2.19: Plot of $h(t)$

2.18 Stability of LTI Systems

In this section, we are interested to answer the following question: if a LTI system is to be BIBO stable (refer Section 2.5), what condition should the impulse resonse satisfy? In other words, if we consider a discrete time LTI system with a bounded input signal $x[n]$ and we want $y[n]$ to be bounded, what condition should $h[n]$ satisfy?

<u>Derivation:</u> Consider a bounded input signal $x[n]$. Thus, $|x[n]| \leq B \forall n$. We want $y[n]$ to be bounded. Since the system is a LTI system,

$$y[n] = x[n] * h[n]$$
$$= h[n] * x[n]$$
$$= \sum_{k=-\infty}^{\infty} h[k]x[n-k]$$
$$|y[n]| = \left| \sum_{k=-\infty}^{\infty} h[k]x[n-k] \right|$$
$$\leq \sum_{k=-\infty}^{\infty} |h[k]x[n-k]| \text{ (because } |a+b| \leq |a| + |b|)$$
$$\leq \sum_{k=-\infty}^{\infty} |h[k]||x[n-k]|$$
$$\leq B \sum_{k=-\infty}^{\infty} |h[k]| \text{ (since } |x[n-k]| \leq B)$$

Thus if we want $y[n]$ to be bounded, the term $\sum_{k=-\infty}^{\infty} |h[k]|$ should also be bounded (i.e. $\sum_{k=-\infty}^{\infty} |h[k]| < 0$). This is the condition on the impulse response if the LTI system is to be BIBO stable.

To summarize,

- For a DT LTI system with impulse response $h[n]$ to be BIBO stable, it must satisfy the following condition:

$$\sum_{k=-\infty}^{\infty} |h[k]| < \infty \qquad (2.25)$$

- Similarly, for a CT LTI system with impulse response $h(t)$ to be BIBO stable, it must satisfy the following condition:

$$\int_{-\infty}^{\infty} |h(\tau)| \, d\tau < \infty \qquad (2.26)$$

2.18.1 Solved Examples

1. A LTI system has an impulse response given by: $h[n] = \alpha^n u[n]$. Is the system stable?

Solution: We need to check if $\sum_{k=-\infty}^{\infty} |h[k]| < \infty$.

$$\sum_{k=-\infty}^{\infty} |h[k]| = \sum_{k=-\infty}^{\infty} |\alpha^k u[k]|$$

$$= \sum_{k=-\infty}^{\infty} |\alpha^k||u[k]|$$

$$= \sum_{k=0}^{\infty} |\alpha|^k$$

This is a geometric series. Thus, if $|\alpha| < 1$,

$$\sum_{k=0}^{\infty} |\alpha|^k = \frac{1}{1-|\alpha|} < \infty,$$

and the system is stable. If $|\alpha| > 1$, the series will not converge and hence the system is not stable.

---

2. Check the stability of the LTI system with impulse response $h(t) = e^{-2t}u(t+2)$.

Solution: We need to check if $\int_{-\infty}^{\infty} |h(\tau)| \, d\tau < \infty$.

$$\int_{-\infty}^{\infty} |h(\tau)| \, d\tau = \int_{-\infty}^{\infty} |e^{-2\tau} u(\tau+2)| \, d\tau$$

$$= \int_{-2}^{\infty} |e^{-2\tau}| \, d\tau$$

$$= \int_{-2}^{\infty} e^{-2\tau} \, d\tau$$

$$= \left[\frac{e^{-2\tau}}{-2} \right]_{-2}^{\infty}$$

$$= \frac{e^4}{2}$$

$$< \infty.$$

Thus the system is stable.

3 Transforms

A transform is basically a mathematical operation that is used to convert a quantity (in the context of this book, a signal) from one domain to another and vice-versa.

This book covers the following six transforms: continuous time Fourier Series and continuous time Fourier Transform (from time domain to continuous time frequency (ω) domain), Laplace Transform (from time domain to s domain), discrete time Fourier Series and discrete time Fourier Transform (from time domain to discrete time frequency (Ω) domain) and the Z Transform (from time domain to z domain).

The type of transform to be performed depends on the type of the signal. This is shown in Table 3.1.

Signal	Periodic signal	Aperiodic signal	
CT (e.g. $x(t)$)	CT Fourier Series	CT Fourier Transform	Laplace Transform
DT (e.g. $x[n]$)	DT Fourier Series	DT Fourier Transform	Z Transform

Table 3.1: The six transforms in this book

The concept map for the topics covered in this chapter is shown in Figure 3.1. In all the transforms that is discussed in this book, the main idea when transforming a signal from one domain to another is to represent the signal as a linear combination of *basis signals*. A detailed discussion on basis signals is not covered here, but in the context of this book it is sufficient to say that the basis signals used in the six transforms presented here, qualify as valid basis signals.

3.1 Continuous Time Fourier Series (CTFS)

In CTFS, we consider a CT periodic signal $x(t)$ with fundamental frequency ω_0 and fundamental period $T = 2\pi/\omega_0$.

The basis signals chosen are complex exponentials (refer Section 1.5) of the form $e^{jk\omega_0 t}$, where ω_0 is the fundamental frequency and $k = 0, \pm 1, \pm 2, \dots$. The basis signals are thus harmonically related with the fundamental frequency ω_0 (meaning,

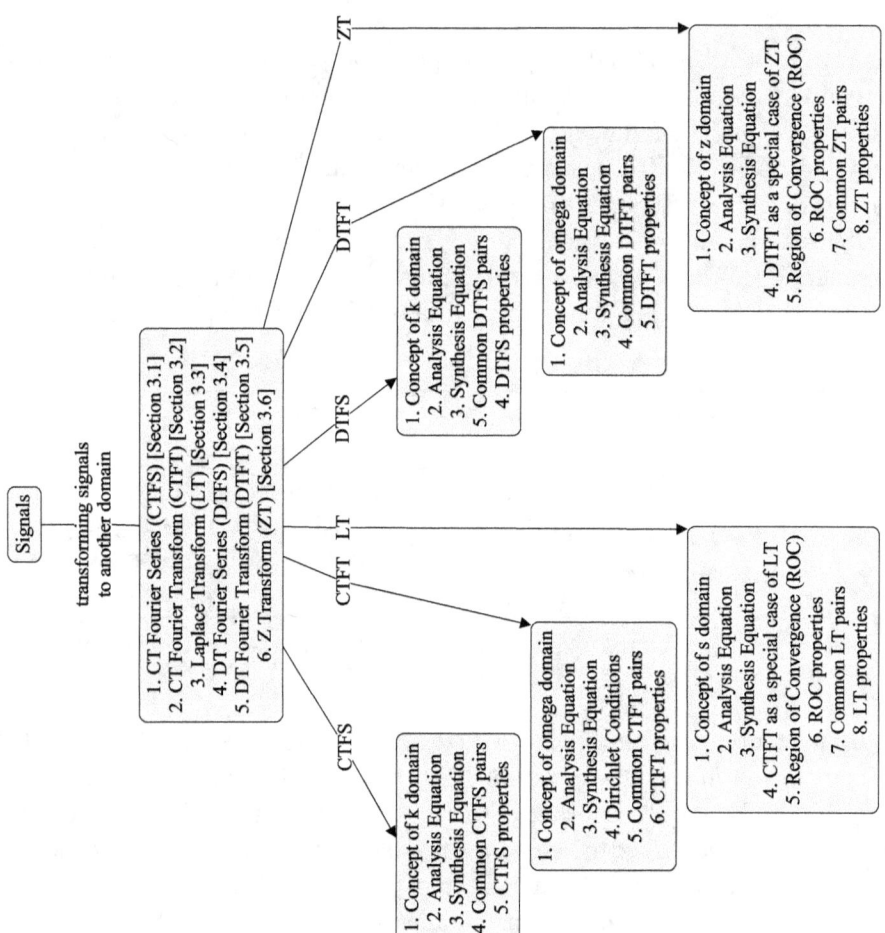

Figure 3.1: Concept map of topics in Chapter 3

kw_0 is the kth harmonic of w_0.). Note that the basis signals are also periodic signals with fundamental frequency w_0.

The time domain signal $x(t)$ can be written as a linear combination of the above basis signals as:

$$x(t) = \sum_{k=-\infty}^{\infty} a_k e^{jkw_0 t}, \qquad (3.1)$$

where a_k are complex constants and are called the Fourier Series coefficients or spectral coefficients of $x(t)$.

Equation (3.1) is the Fourier Series representation of the periodic signal $x(t)$. Another way of looking at Equation (3.1) is, if we know w_0, k (integers from $-\infty$ to $+\infty$) and a_k (for each value of k), meaning the right hand side of Equation (3.1)), then we can compute the time domain signal $x(t)$. In this sense we are transforming the signal from the w (frequency domain, or equivalently k domain) to the time domain. This is shown in Figure 3.2.

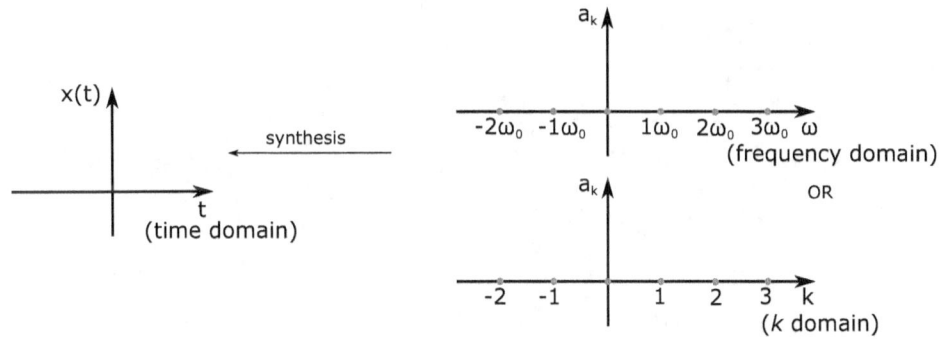

Figure 3.2: CTFS synthesis equation

a_k can be derived from Equation (3.1), but its derivation is beyond the scope this book. The expression for a_k can be derived to be:

$$a_k = \frac{1}{T} \int_T x(t) e^{-jkw_0 t} \, dt. \qquad (3.2)$$

Another way looking at Equation (3.2) is, if we know w_0, k (integers from $-\infty$ to $+\infty$) and $x(t)$, meaning the right hand side of Equation (3.2)), then we can compute a_k (for each value of k). In this sense we are transforming the signal from the time domain to the w (frequency domain, or equivalently k domain). This is shown in Figure 3.3.

To summarize, the two equations for CTFS (called the analysis and synthesis equations) are given by:

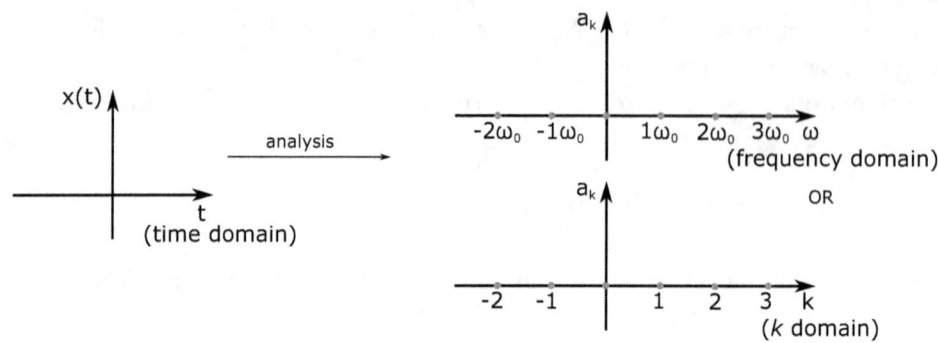

Figure 3.3: CTFS analysis equation

Synthesis equation:
$$x(t) = \sum_{k=-\infty}^{\infty} a_k e^{jk\omega_0 t} \tag{3.3}$$

Analysis equation:
$$a_k = \frac{1}{T} \int_T x(t) e^{-jk\omega_0 t} \, dt \tag{3.4}$$

Transform represented by:
$$x(t) \xleftrightarrow{\text{CTFS}} a_k$$

There are two methods to solve numerical problems related to finding the CTFS coefficients a_k. The first method, a comparision (to Equation (3.3)) method is used for problems where the time-domain signal is trigonometric function. For other types of signals, the second method of directly applying Equation (3.4) is used. Both these methods are demonstrated in the examples below.

3.1.1 Solved Examples

1. $x(t) = \sin(\omega_0 t)$. Find the Fourier Series coefficients.

Solution: $x(t)$ is a periodic signal with fundamental frequency ω_0. Since $x(t)$ is a trigonometric function, we will use the comparision method. To do so, we must first expand $x(t)$ in terms of its fundamental frequency.

80

$$x(t) = \sin(\omega_0 t)$$
$$= \frac{e^{j\omega_0 t} - e^{-j\omega_0 t}}{2j}$$
$$= \frac{1}{2j}e^{j\omega_0 t} - \frac{1}{2j}e^{-j\omega_0 t}$$

Comparing with Equation (3.3), we can write

$$a_1(\text{when } k = 1) = \frac{1}{2j}, a_{-1}(\text{when } k = -1) = \frac{-1}{2j}.$$

Note that a_k is a complex number. The plot of the magnitude and phase of a_k are shown in Figure 3.4a and 3.4b.

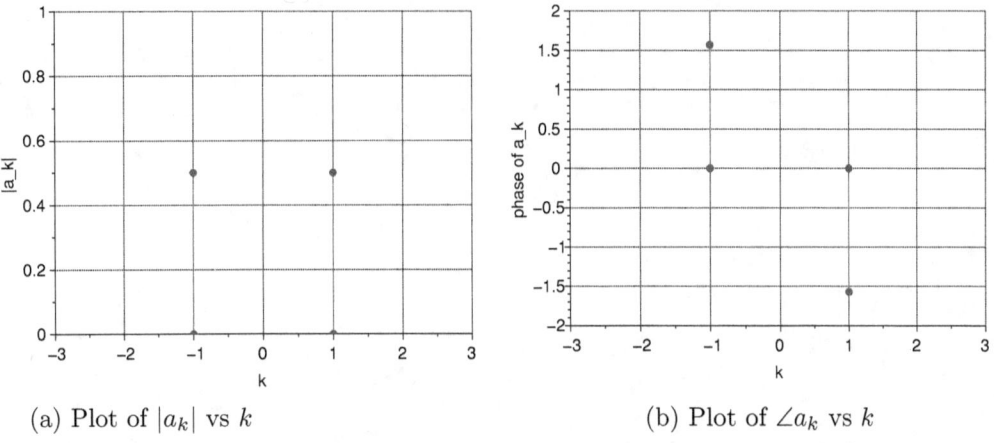

(a) Plot of $|a_k|$ vs k (b) Plot of $\angle a_k$ vs k

Figure 3.4: Plot of a_k

2. Consider a signal $x(t)$ shown in Figure 3.5, with $T > T_1$. It is defined as follows:
$$x(t) = \begin{cases} 1, & |t| < T_1 \\ 0, & T_1 < t < T/2 \\ 0, & -T/2 < t < -T_1 \end{cases}$$
Find the Fourier Series representation of $x(t)$.

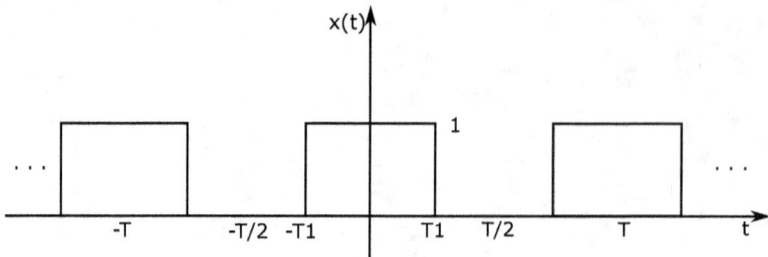

Figure 3.5: A periodic signal $x(t)$ (for solved example in Section 3.1.1)

Solution: $x(t)$ is a periodic signal with period T. To find the CTFS coefficients we will directly apply the analysis equation when $k = 0$ and when $k \neq 0$.

When $k = 0$,

$$
\begin{aligned}
a_0 &= \frac{1}{T} \int_{<T>} x(t) \, dt \\
&= \frac{1}{T} \int_{-T/2}^{T/2} x(t) \, dt \quad \text{(this interval makes the mathematics easier)} \\
&= \frac{1}{T} \int_{-T_1}^{T_1} x(t) \, dt \quad \text{(since } x(t) = 0 \text{ outside this interval)} \\
&= \frac{1}{T} \int_{-T_1}^{T_1} dt \quad \text{(since } x(t) = 1 \text{ inside this interval)} \\
&= \frac{2T_1}{T}.
\end{aligned}
$$

When $k \neq 0$

$$
\begin{aligned}
a_k &= \frac{1}{T} \int_T x(t) e^{-jk\omega_0 t} \, dt \\
&= \frac{1}{T} \int_{-T}^{T} x(t) e^{-jk\omega_0 t} \, dt \\
&= \frac{1}{T} \int_{-T_1}^{T_1} e^{-jk\omega_0 t} \, dt \\
&= \frac{1}{T} \left[\frac{e^{-jk\omega_0 t}}{-jk\omega_0} \right]_{T_1}^{T_1} \\
&= \frac{1}{-jk\omega_0 T} \left[e^{-jk\omega_0 T_1} - e^{jk\omega_0 T_1} \right] \\
&= \frac{2}{k\omega_0 T} \left[\frac{e^{jk\omega_0 T_1} - e^{-jk\omega_0 T_1}}{2j} \right] \\
&= \frac{1}{k\pi} \sin(k\omega_0 T_1). \quad (\text{since } \omega_0 T = 2\pi)
\end{aligned}
$$

Therefore,

$$
a_k = \begin{cases}
\frac{2T_1}{T} & , k = 0 \\[2em]
\frac{1}{k\pi} \sin(k\omega_0 T_1) & , k \neq 0.
\end{cases}
$$

---

3. $x(t) = 2 + 3\cos(2\pi t) + 4\sin(3\pi t)$. Find its Fourier Series coefficients.

Solution: First, we need to check if $x(t)$ is a periodic signal. Since $x(t)$ is a sum of two sinusoidal signals, we will use the approach seen earlier in Section 1.10.

For $\cos(2\pi t)$,

$$
\cos(2\pi t) \Rightarrow \omega_1 = 2\pi
$$
$$
T_1 = \frac{2\pi}{\omega_1} = 1
$$

For $\sin(3\pi t)$,

$$
\sin(3\pi t) \Rightarrow \omega_2 = 3\pi
$$
$$
T_2 = \frac{2\pi}{\omega_2} = \frac{2}{3}
$$

Therefore,

$$\frac{T_1}{T_2} = \frac{1}{\frac{2}{3}} = 3/2.$$

Since $\frac{T_1}{T_2}$ is a rational number, the signal $x(t)$ is periodic. The periodicity is given by $\text{LCM}(1, \frac{3}{2}) = T_0 = 2$ [sec]. The fundamental frequency is given by $\omega_0 = \frac{2\pi}{T_0} = \frac{2\pi}{2} = \pi$. As done earlier, since $x(t)$ is a trigonometric function, we will be using the comparision method to calculate the CTFS coefficients.

Writing $x(t)$ in terms of ω_0 gives,

$$x(t) = 2 + 3\cos(2\omega_0 t) + 4\sin(3\omega_0 t)$$
$$= 2 + 3\left(\frac{e^{j2\omega_0 t} + e^{-j2\omega_0 t}}{2}\right) + 4\left(\frac{e^{j3\omega_0 t} - e^{-j3\omega_0 t}}{2j}\right)$$
$$= 2 + \frac{3}{2}e^{j2\omega_0 t} + \frac{3}{2}e^{-j2\omega_0 t} + \frac{2}{j}e^{j3\omega_0 t} - \frac{2}{j}e^{-j3\omega_0 t}$$

Comparing with Equation (3.3), the CTFS coefficients are:

$$a_0 = 2, a_2 = \frac{3}{2}, a_{-2} = \frac{3}{2}, a_3 = \frac{2}{j}, a_{-3} = \frac{2}{j}.$$

---

4. Consider the signal $w(t) = A$, where A is a constant. Determine the Fourier Series expansion of $w(t)$.

Solution: $x(t)$ is a periodic signal with period T. To find the CTFS coefficients we will directly apply the analysis equation when $k = 0$ and when $k \neq 0$.

When $k = 0$,

$$a_0 = \frac{1}{T}\int_{<T>} x(t)\, dt$$
$$= \frac{1}{T}\int_0^T x(t)\, dt$$
$$= \frac{1}{T}\int_0^T A\, dt$$
$$= A.$$

When $k \neq 0$

$$
\begin{aligned}
a_k &= \frac{1}{T} \int_T x(t) e^{-jk\omega_0 t} \, dt \\
&= \frac{1}{T} \int_0^T A e^{-jk\omega_0 t} \, dt \\
&= \frac{A}{T} \int_0^T e^{-jk\omega_0 t} \, dt \\
&= \frac{A}{T} \left[\frac{e^{-jk\omega_0 t}}{-jk\omega_0} \right]_0^T \\
&= \frac{A}{-jk\omega_0 T} \left[e^{-jk\omega_0 T} - e^{j0} \right] \\
&= \frac{A}{-jk\omega_0 T} \left[e^{-jk2\pi} - e^{j0} \right] \quad \text{since } \omega_0 T = 2\pi \\
&= \frac{A}{-jk\omega_0 T} [1 - 1] \\
&= 0.
\end{aligned}
\tag{3.5}
$$

Therefore,

$$
a_k = \begin{cases} A & , k = 0 \\ \\ 0 & , k \neq 0. \end{cases}
$$

3.1.2 Properties of CTFS

Consider a periodic signal $x(t)$ with Fourier Series coefficients a_k, that is,

$$
x(t) \xleftrightarrow{\text{CTFS}} a_k.
$$

We are interested to know what happens to the Fourier Series coefficients or what happens in the k domain when various signal operations are done in the time (t) domain.

Some of the major properties of the CTFS that are often used when solving numberical problems are discussed next; for a full list of the properties please refer [1].

Linearity

If $x(t)$ and $y(t)$ are periodic signals with period T and

$$
\begin{aligned}
x(t) &\xleftrightarrow{\text{CTFS}} a_k \\
y(t) &\xleftrightarrow{\text{CTFS}} b_k
\end{aligned}
$$

Then
$$z(t) = Ax(t) + By(t) \xleftrightarrow{\text{CTFS}} c_k = Aa_k + Bb_k, \qquad (3.6)$$

where c_k are the Fourier Series coefficients of $z(t)$.

Therefore, *linearity in the time domain* corresponds to *linearity in the k domain*.

Time Shifting

If $x(t)$ is a periodic signal with period T and

$$x(t) \xleftrightarrow{\text{CTFS}} a_k$$

Then

$$x(t - t_0) \xleftrightarrow{\text{CTFS}} b_k = e^{-jk\omega_0 t_0} a_k \qquad (3.7)$$

where b_k are the Fourier Series coefficients of $x(t - t_0)$. Note that $x(t - t_0)$ is also periodic in T.

Therefore, *shifting in the time domain* corresponds to *phase shift in the k domain*. Magnitude spectrum of a_k and b_k will remain the same.

Time Reversal

If $x(t)$ is a periodic signal with period T and

$$x(t) \xleftrightarrow{\text{CTFS}} a_k$$

Then

$$x(-t) \xleftrightarrow{\text{CTFS}} b_k = a_{-k} \qquad (3.8)$$

where b_k are the Fourier Series coefficients of $x(-t)$. Note that $x(-t)$ is also periodic in T.

Therefore, *reversal in the time domain* corresponds to *reversal in the k domain*.

Time Scaling

If $x(t)$ is a periodic signal with period T and

$$x(t) \xleftrightarrow{\text{CTFS}} a_k$$

Then

$$x(\alpha t), \alpha > 0 \xleftrightarrow{\text{CTFS}} b_k = a_k \qquad (3.9)$$

where b_k are the Fourier Series coefficients of $x(\alpha t)$. Note that $x(\alpha t)$ is periodic in T/α.

Multiplication

If $x(t)$ and $y(t)$ are periodic signals with period T and

$$x(t) \xleftrightarrow{\text{CTFS}} a_k$$
$$y(t) \xleftrightarrow{\text{CTFS}} b_k$$

Then

$$z(t) = x(t)y(t) \xleftrightarrow{\text{CTFS}} c_k = \sum_{l=-\infty}^{\infty} a_l b_{k-l} \tag{3.10}$$

where c_k are the Fourier Series coefficients of $z(t)$.

Therefore, *multiplication in the time domain* corresponds to *convolution in the k domain*.

Convolution

If $x(t)$ and $y(t)$ are periodic signals with period T and

$$x(t) \xleftrightarrow{\text{CTFS}} a_k$$
$$y(t) \xleftrightarrow{\text{CTFS}} b_k$$

Then

$$z(t) = x(t) * y(t) \xleftrightarrow{\text{CTFS}} c_k = T a_k b_k \tag{3.11}$$

where c_k are the Fourier Series coefficients of $z(t)$.

Therefore, *convolution in the time domain* corresponds to *multiplication in the k domain*.

Conjugation

If $x(t)$ is a periodic signal with period T and

$$x(t) \xleftrightarrow{\text{CTFS}} a_k$$

Then

$$x^*(t) \xleftrightarrow{\text{CTFS}} b_k = a^*_{-k} \tag{3.12}$$

where b_k are the Fourier Series coefficients of $x^*(t)$. Note that $x^*(t)$ is periodic in T.

Frequency Shift

If $x(t)$ is a periodic signal with period T and

$$x(t) \xleftrightarrow{\text{CTFS}} a_k$$

Then

$$e^{jM\omega_0 t} x(t) \xleftrightarrow{\text{CTFS}} b_k = a_{k-M} \tag{3.13}$$

where b_k are the Fourier Series coefficients of $e^{jM\omega_0 t} x(t)$.

Therefore, *phase shift in the time domain* corresponds to *frequency shift in the k domain*.

Parseval's Theorem

If $x(t)$ is a periodic signal with period T and

$$x(t) \xleftrightarrow{\text{CTFS}} a_k$$

Then

$$\frac{1}{T} \int_T |x(t)|^2 dt = \sum_{k=-\infty}^{\infty} |a_k|^2 \tag{3.14}$$

3.1.3 Solved Example

1. Consider the signal shown in Figure 3.5. Let $T_1 = 1$ and $T = 4$. Draw the signal $g(t) = x(t-1) - 1/2$. Find the Fourier Series expansion of $g(t)$.

Solution: To get $g(t)$, we first shift $x(t)$ to the right by $t_0 = 1$. Then a downward amplitude shift of $1/2$ is applied. The resultant signal $g(t)$ is shown in Figure 3.6.

Let a_k by the Fourier Series coefficients of $x(t)$. a_k is given by (see Section 3.1.1):

$$a_k = \begin{cases} \frac{2T_1}{T} & , k = 0 \\ \\ \frac{1}{k\pi} \sin(k\omega_0 T_1) & , k \neq 0 \end{cases}$$

Let b_k be the Fourier Series coefficients of the signal $x(t-1)$. Applying the time-shifting property of the CTFS,

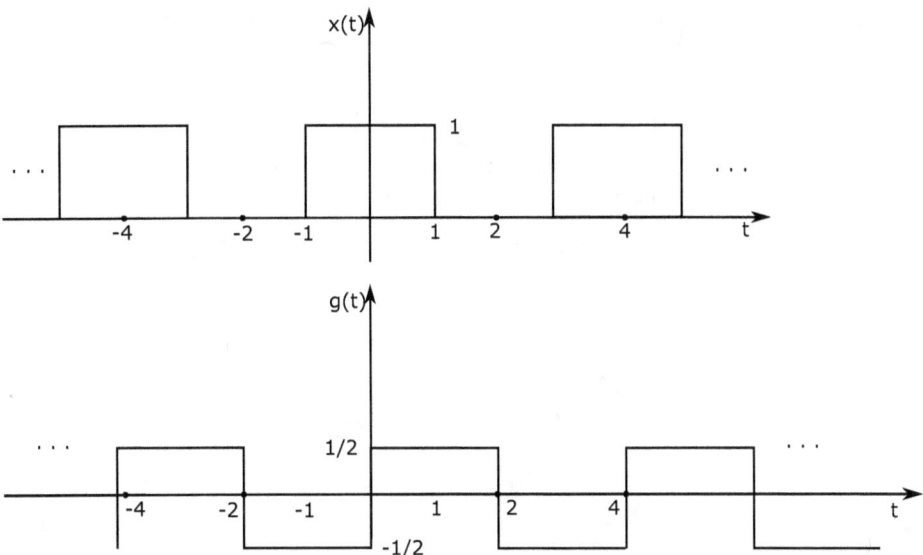

Figure 3.6: A periodic signal $g(t)$ (for solved example in Section 3.1.3)

$$x(t-1) \xleftrightarrow{\text{CTFS}} b_k = a_k e^{-jk\omega_0 t_0}$$
$$b_k = a_k e^{-jk\omega_0} \quad (\text{since } t_0 = 1).$$

Therefore

$$b_k = \begin{cases} \frac{2T_1}{T} e^{-jk\omega_0} & , k = 0 \\ \\ \frac{1}{k\pi} \sin(k\omega_0 T_1) e^{-jk\omega_0} & , k \neq 0 \end{cases}$$

Substituting values for T_1 and T,

$$b_k = \begin{cases} \frac{1}{2} & , k = 0 \\ \\ \frac{1}{k\pi} \sin(k\omega_0) e^{-jk\omega_0} & , k \neq 0 \end{cases}$$

Let c_k be the Fourier Series coefficients of the signal $g(t) = x(t-1) - 1/2$. To calculate c_k we need to first know the CTFS coefficients of a constant signal $z(t) = A$ (in this problem $A = -1/2$).

$$z(t) = A \xleftrightarrow{\text{CTFS}} \begin{cases} A & , k = 0 \\ 0 & , k \neq 0 \end{cases}$$

Now, applying the linearity property of the CTFS (since $g(t) = x(t-1) + z(t)$), we get

$$c_k = \begin{cases} \frac{1}{2} + \left(\frac{-1}{2}\right) & , k = 0 \\ \frac{1}{k\pi}\sin(k\omega_0)e^{-jk\omega_0} + 0 & , k \neq 0 \end{cases}$$

Therefore,

$$c_k = \begin{cases} 0 & , k = 0 \\ \frac{1}{k\pi}\sin(k\omega_0)e^{-jk\omega_0} & , k \neq 0 \end{cases}$$

where $\omega_0 = \frac{2\pi}{T} = \frac{\pi}{2}$.

3.2 Continuous Time Fourier Transform (CTFT)

In CTFT, we consider a CT aperiodic signal $x(t)$. We are going to represent $x(t)$ as a linear combination of basis signals. The basis signals chosen are complex exponentials (refer Section 1.5) of the form $e^{j\omega t}$.

The time domain signal $x(t)$ can be written as a linear combination of the above basis signals as:

$$x(t) = \frac{1}{2\pi}\int_{-\infty}^{\infty} X(j\omega)e^{j\omega t}d\omega \tag{3.15}$$

where $X(j\omega)$ is called the Fourier Transform or spectrum of $x(t)$.

Another way of looking at Equation (3.15) is, if we know ω and $X(j\omega)$ (maybe from experimental measurements or simulation), meaning the right hand side of Equation (3.15)), then we can compute the time domain signal $x(t)$. In this sense we are transforming the signal from the ω (frequency) domain to the time domain. This is shown in Figure 3.7.

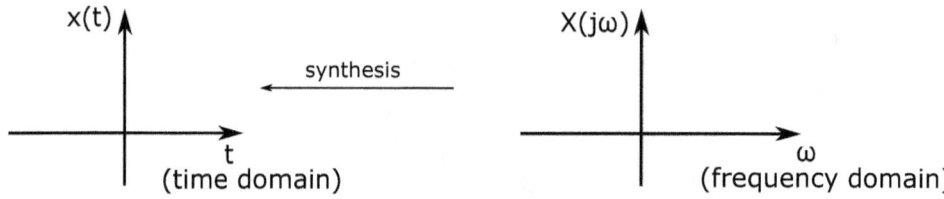

Figure 3.7: CTFT synthesis equation

The derivation of $X(j\omega)$ is beyond the scope of this book, and those interested can refer Section 4.1 in [1]. The expression for $X(j\omega)$ is given by:

$$X(j\omega) = \int_{-\infty}^{\infty} x(t)e^{-j\omega t}dt. \tag{3.16}$$

Another way looking at Equation (3.16) is, if we know ω and $x(t)$, meaning the right hand side of Equation (3.16)), then we can compute $X(j\omega)$ (for each value of ω). In this sense we are transforming the signal from the time domain to the ω (frequency) domain. This is shown in Figure 3.8.

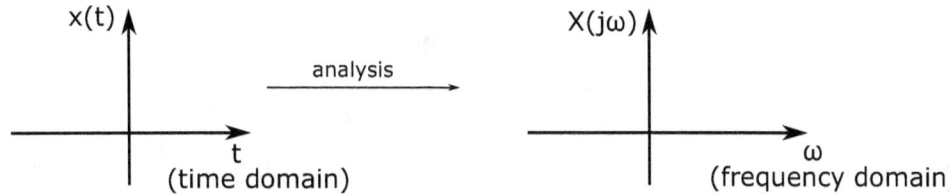

Figure 3.8: CTFT analysis equation

Note that when plotting the Fourier Transform, the horizontal axis is ω, which is a continuous variable. Thus, the CTFT plot is a continuous domain/frequency plot (on the other hand, the CTFS plot is a discrete domain/frequency plot).

To summarize, the two equations for CTFT (called the analysis and synthesis equations) are given by:

Synthesis equation:

$$x(t) = \frac{1}{2\pi} \int_{-\infty}^{\infty} X(j\omega)e^{j\omega t}d\omega \tag{3.17}$$

Analysis equation:

$$X(j\omega) = \int_{-\infty}^{\infty} x(t)e^{-j\omega t}dt \tag{3.18}$$

Transform represented by:

$$x(t) \xleftrightarrow{\text{CTFT}} X(j\omega)$$

Dirichlet Conditions

The sufficient conditions for the convergence of the Fourier Transform ($X(j\omega)$) are the following (also referred to as the Dirichlet conditions):

- $x(t)$ is absolutely integrable, that is

$$\int_{-\infty}^{\infty} |x(t)|dt < \infty \qquad (3.19)$$

- $x(t)$ has a finite number of maxima and minima within any finite interval.

- $x(t)$ has a finite number of discontinuties within any finite interval, and each of these discontinuities is finite.

3.2.1 Common Fourier Transform Pairs

Some commonly used CTFT pairs are given in Table 3.2.

Signal	CT Fourier Transform
$x(t) = 1$	$2\pi\delta(\omega)$
$\delta(t)$	1
$u(t)$	$\frac{1}{j\omega} + \pi\delta(\omega)$
$\cos(\omega_0 t)$	$\pi[\delta(\omega - \omega_0) + \delta(\omega + \omega_0)]$
$\sin(\omega_0 t)$	$\frac{\pi}{j}[\delta(\omega - \omega_0) + \delta(\omega + \omega_0)]$
$e^{-at}u(t), Re(a) > 0$	$\frac{1}{a+j\omega}$
$te^{-at}u(t), Re(a) > 0$	$\frac{1}{(a+j\omega)^2}$

Table 3.2: Common CTFT pairs

3.2.2 Properties of the CTFT

Consider a signal $x(t)$ with Fourier coefficients $X(j\omega)$, that is,

$$x(t) \xleftrightarrow{\text{CTFT}} X(j\omega).$$

We are interested to know what happens to the Fourier Transform coefficients when various signal operations are done in the time (t) domain.

Some of the major properties of the CTFT that are often used when solving numberical problems are discussed next; for a full list of the properties please refer [1].

Linearity

Consider two signals $x(t)$ and $y(t)$ with

$$x(t) \xleftrightarrow{\text{CTFT}} X(j\omega)$$
$$y(t) \xleftrightarrow{\text{CTFT}} Y(j\omega)$$

Then

$$z(t) = Ax(t) + By(t) \xleftrightarrow{\text{CTFT}} Z(j\omega) = AX(j\omega) + BY(j\omega), \qquad (3.20)$$

where $Z(j\omega)$ are the Fourier Transform coefficients of $z(t)$.

Therefore, *linearity in the time domain* corresponds to *linearity in the ω domain*.

Time Shifting

Consider $x(t)$ with

$$x(t) \xleftrightarrow{\text{CTFT}} X(j\omega)$$

Then

$$x(t - t_0) \xleftrightarrow{\text{CTFT}} e^{-j\omega t_0} X(j\omega). \qquad (3.21)$$

Therefore, *shifting in the time domain* corresponds to *phase shift in the ω domain*.

Time Reversal

Consider $x(t)$ with

$$x(t) \xleftrightarrow{\text{CTFT}} X(j\omega)$$

Then

$$x(-t) \xleftrightarrow{\text{CTFT}} X(-j\omega). \qquad (3.22)$$

Therefore, *reversal in the time domain* corresponds to *reversal in the ω domain*.

Time Scaling

Consider $x(t)$ with

$$x(t) \xleftrightarrow{\text{CTFT}} X(j\omega)$$

Then

$$x(\alpha t), \alpha > 0 \xleftrightarrow{\text{CTFT}} \frac{1}{|\alpha|} X(\frac{j\omega}{\alpha}). \qquad (3.23)$$

Multiplication

Consider $x(t)$ and $y(t)$ with

$$x(t) \xleftrightarrow{\text{CTFT}} X(j\omega)$$
$$y(t) \xleftrightarrow{\text{CTFT}} Y(j\omega)$$

Then

$$z(t) = x(t)y(t) \xleftrightarrow{\text{CTFT}} \frac{1}{2\pi} X(j\omega) * Y(j\omega) = \frac{1}{2\pi} \int_{-\infty}^{\infty} X(j\theta)Y(j(\omega - \theta))\, d\theta \quad (3.24)$$

Therefore, *multiplication in the time domain* corresponds to *convolution in the ω domain*.

Convolution

Consider $x(t)$ and $y(t)$ with

$$x(t) \xleftrightarrow{\text{CTFT}} X(j\omega)$$
$$y(t) \xleftrightarrow{\text{CTFT}} Y(j\omega)$$

Then

$$z(t) = x(t) * y(t) \xleftrightarrow{\text{CTFT}} X(j\omega)Y(j\omega). \quad (3.25)$$

Therefore, *convolution in the time domain* corresponds to *multiplication in the ω domain*.

Conjugation

Consider $x(t)$ with

$$x(t) \xleftrightarrow{\text{CTFT}} X(j\omega)$$

Then

$$x^*(t) \xleftrightarrow{\text{CTFT}} X^*(-j\omega). \quad (3.26)$$

Frequency Shift

Consider $x(t)$ with

$$x(t) \xleftrightarrow{\text{CTFT}} X(j\omega)$$

Then

$$e^{j\omega_0 t} x(t) \xleftrightarrow{\text{CTFT}} X(j(\omega - \omega_0)). \quad (3.27)$$

Therefore, *phase shift in the time domain* corresponds to *frequency shift in the ω domain*.

Differentiation in Time

Consider $x(t)$ with

$$x(t) \xleftrightarrow{\text{CTFT}} X(j\omega)$$

Then

$$\frac{d}{dt}x(t) \xleftrightarrow{\text{CTFT}} j\omega X(j\omega). \tag{3.28}$$

Integration in Time

Consider $x(t)$ with

$$x(t) \xleftrightarrow{\text{CTFT}} X(j\omega)$$

Then

$$\int_{-\infty}^{t} x(t)\, dt \xleftrightarrow{\text{CTFT}} \frac{1}{j\omega}X(j\omega) + \pi X(0)\delta(\omega) \tag{3.29}$$

Parseval's Theorem

Consider $x(t)$ with

$$x(t) \xleftrightarrow{\text{CTFT}} X(j\omega)$$

Then

$$\int_{-\infty}^{\infty} |x(t)|^2 dt = \frac{1}{2\pi} \int_{-\infty}^{\infty} |X(j\omega)|^2\, d\omega \tag{3.30}$$

3.2.3 Solved Examples

---

1. $x(t) = \delta(t)$. Find the Fourier Transform.

Solution: Applying the analysis equation,

$$\begin{aligned}
X(j\omega) &= \int_{-\infty}^{\infty} \delta(t)e^{-j\omega t}\, dt \\
&= \left[e^{-j\omega t}\right]_{t=0} \text{ (using Equation (1.5))} \\
&= 1. \tag{3.31}
\end{aligned}$$

---

2. $x(t) = e^{-at}u(t), a > 0$ and a is a real number. Find the Fourier Transform.

95

Solution: Applying the analysis equation,

$$X(j\omega) = \int_{-\infty}^{\infty} x(t)e^{-j\omega t}\, dt$$

$$= \int_{-\infty}^{\infty} e^{-at}u(t)e^{-j\omega t}\, dt$$

$$= \int_{0}^{\infty} e^{-at}e^{-j\omega t}\, dt$$

$$= \int_{-\infty}^{\infty} e^{-(a+j\omega)t}\, dt$$

$$= \frac{-1}{a+j\omega}\left[e^{-(a+j\omega)t}\right]_{0}^{\infty}$$

$$= \frac{-1}{a+j\omega}\left[e^{-\infty} - 1\right]$$

$$= \frac{1}{a+j\omega}. \quad (a > 0)$$

---

3. Consider the rectangular pulse signal

$$x(t) = \begin{cases} 1, & |t| < T_1 \\ 0, & |t| > T_1 \end{cases}$$

Find the Fourier Transform of $x(t)$.

Solution: Applying the analysis equation,

$$X(j\omega) = \int_{-\infty}^{\infty} x(t)e^{-j\omega t}\, dt$$

$$= \int_{-T_1}^{T_1} e^{-j\omega t}\, dt \quad (\text{because} - T_1 < x(t) = 1 < T_1)$$

$$= \frac{-1}{j\omega}\left[e^{-j\omega t}\right]_{-T_1}^{T_1}$$

$$= \frac{-1}{j\omega}\left[e^{-j\omega T_1} - e^{j\omega T_1}\right]$$

$$= \frac{2}{\omega}\sin(\omega T_1). \text{ (multiply and divide by 2, then rearrange)}$$

---

4. $x(t) = u(t)$. Find the Fourier Transform.

Solution: From Equation (1.11), $u(t)$ and $\delta(t)$ are related by:

$$x(t) = u(t) = \int_{\tau=-\infty}^{t} \delta(\tau)d\tau = \int_{\tau=-\infty}^{t} g(\tau)d\tau$$

where

$$g(t) = \delta(t) \xleftrightarrow{\text{CTFT}} G(j\omega) = 1.$$

Using the integration property of the CTFT,

$$x(t) \xleftrightarrow{\text{CTFT}} X(j\omega) = \frac{G(j\omega)}{j\omega} + jG(0)\delta(\omega)$$

$$= \frac{1}{j\omega} + j\delta(\omega).$$

---

5. Recall from the properties of the CTFT that, multiplication in the time domain leads to convolution in the frequency domain. Given $s(t)$ with $S(j\omega)$ as shown in Figure 3.9; $p(t) = \cos(\omega_0 t)$ and $r(t) = s(t)p(t)$. Calculate $R(j\omega)$.

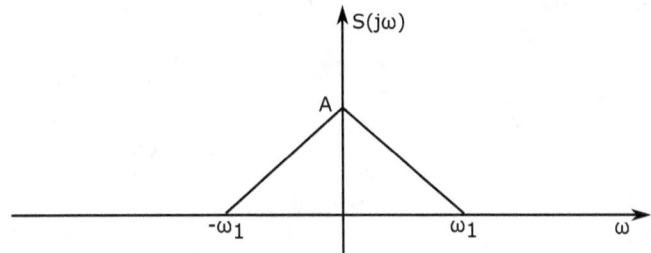

Figure 3.9: $S(j\omega)$

Solution: From the multiplication property of the CTFT,

$$R(j\omega) = \frac{1}{2\pi} S(j\omega) * P(j\omega).$$

$S(j\omega)$ is given in Figure 3.9. From Table 3.2, the CTFT of $\cos(\omega_0 t)$ is given by

$$P(j\omega) = \pi\delta(\omega - \omega_0) + \pi\delta(\omega + \omega_0).$$

This is shown in Figure 3.10.

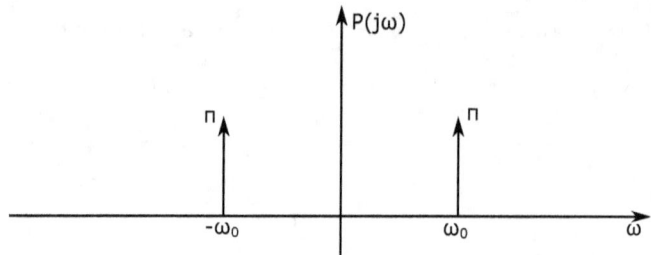

Figure 3.10: $P(j\omega)$

Thus $R(j\omega)$ is

$$R(j\omega) = \frac{1}{2\pi} S(j\omega) * P(j\omega)$$

$$= \frac{1}{2\pi} \int_{-\infty}^{\infty} S(j\theta)P(j(\omega - \theta))d\theta$$

$$= \frac{1}{2\pi} \int_{-\infty}^{\infty} S(j\theta) \left[\pi\delta(\omega - \omega_0) + \pi\delta(\omega + \omega_0) \right] d\theta$$

$$= \frac{1}{2} \int_{-\infty}^{\infty} S(j\theta)\delta(\omega - \omega_0 - \theta)d\theta + \frac{1}{2} \int_{-\infty}^{\infty} S(j\theta)\delta(\omega + \omega_0 - \theta)d\theta$$

From the sampling property of the impulse function

$$\int_{-\infty}^{\infty} S(j\theta)\delta(\omega - \omega_0 - \theta)d\theta = S(j(\omega - \omega_0))$$

$$\int_{-\infty}^{\infty} S(j\theta)\delta(\omega + \omega_0 - \theta)d\theta = S(j(\omega + \omega_0))$$

Thus $R(j\omega)$ becomes,

$$R(j\omega) = \frac{1}{2}S(j(\omega - \omega_0)) + \frac{1}{2}S(j(\omega + \omega_0)).$$

This is shown in Figure 3.11.

Thus when we convolve any signal with the δ signal, we get copies of the original signal at the location of the impulses.

3.3 Laplace Transform (LT)

Consider a continuous time signal $x(t)$. We are going to represent $x(t)$ as a linear combination of basis signals. The basis signals chosen are complex exponentials (refer Section 1.5) of the form e^{st}, where $s = \sigma + j\omega$ is a complex number.

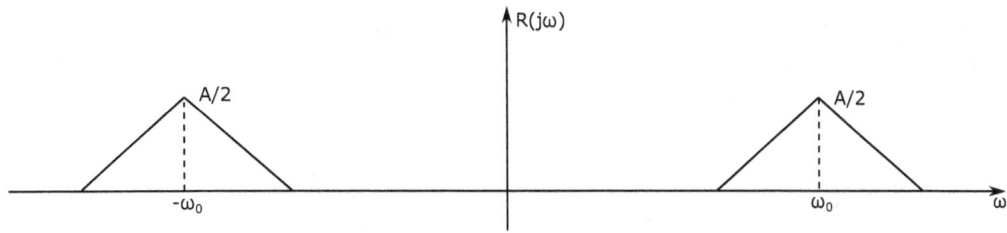

Figure 3.11: $R(j\omega)$

The time domain signal $x(t)$ can be written as a linear combination of the above basis signals as:

$$x(t) = \frac{1}{2\pi} \int_{-\infty}^{\infty} X(\sigma + j\omega)e^{(\sigma+j\omega)t}d\omega \qquad (3.32)$$

In the complex domain, the above equation can be written as:

$$x(t) = \frac{1}{2\pi j} \int_{\sigma-j\omega}^{\sigma+j\omega} X(s)e^{st}ds \qquad (3.33)$$

Note that $X(s)$ is called the Laplace Transform of $x(t)$.

Another way of looking at Equation (3.33) is, if we know s and $X(s)$, meaning the right hand side of Equation (3.33)), then we can compute the time domain signal $x(t)$. In this sense we are transforming the signal from the complex s domain to the time domain. This is shown in Figure 3.12. Note that for the numerical problems in this book, both Equations (3.32) and (3.33) are not used. Instead, the partial-fraction expansion method is used.

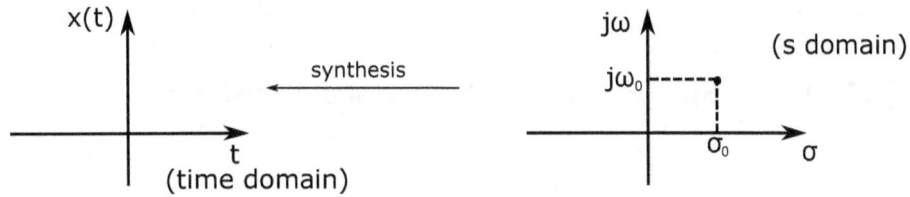

Figure 3.12: Laplace Transform synthesis equation

The expression for $X(s)$ is given by:

$$X(s) = X(\sigma + j\omega) = \int_{-\infty}^{\infty} x(t)e^{-st}dt. \qquad (3.34)$$

Another way looking at Equation (3.34) is, if we know s and $x(t)$, meaning the right hand side of Equation (3.34)), then we can compute $X(s)$ (for each value of s).

In this sense we are transforming the signal from the time domain to the complex s domain. This is shown in Figure 3.13. Equation (3.34) is called the Bilateral Laplace Transform of $x(t)$.

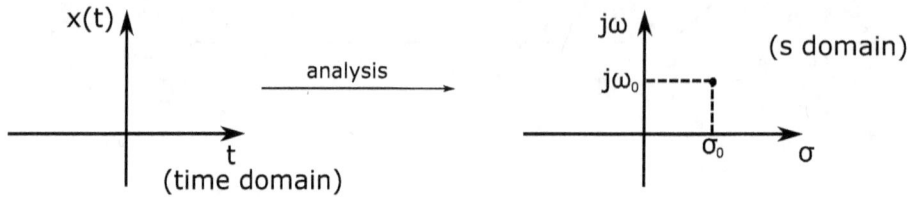

Figure 3.13: Laplace Transform analysis equation

To summarize, the two equations for the Bilateral Laplace Transform (called the analysis and synthesis equations) are given by:

Synthesis equation:

$$x(t) = \frac{1}{2\pi j} \int_{\sigma-j\omega}^{\sigma+j\omega} X(s)e^{st}ds \qquad (3.35)$$

Analysis equation:

$$X(s) = X(\sigma + j\omega) = \int_{-\infty}^{\infty} x(t)e^{-st}dt \qquad (3.36)$$

Transform represented by:

$$x(t) \xleftrightarrow{\text{LT}} X(s)$$

3.3.1 CTFT as a Special Case of the Laplace Transform

From Equation (3.36),

$$X(s) = \int_{-\infty}^{\infty} x(t)e^{-(\sigma+j\omega)t}dt$$
$$= \int_{-\infty}^{\infty} x(t)e^{-\sigma t}e^{-j\omega t}dt$$
$$= \text{CTFT}\left\{x(t)e^{-\sigma t}\right\}$$

If $\sigma = 0$, that is $s = j\omega$, then, $X(s) = \text{CTFT}\{x(t)\}$, that is the Laplace Transform of $x(t)$ is the same as the CTFT of $x(t)$.

3.3.2 Region of Convergence (ROC)

The range of values of s for which Equation (3.36) converges is called the ROC of the Laplace Transform. Note that s is a complex number.

3.3.3 Solved Examples

1. Find the Laplace Transform of $x(t) = e^{-at}u(t), a > 0, a$ is a real number.

Solution: Using the analysis equation (Equation (3.36)),

$$X(s) = \int_{-\infty}^{\infty} x(t)e^{-st}dt$$

$$= \int_{-\infty}^{\infty} e^{-at}u(t)e^{-st}dt$$

$$= \int_{-\infty}^{\infty} e^{-at}e^{-\sigma t}u(t)e^{-j\omega t}dt$$

$$= \int_{-\infty}^{\infty} e^{-(a+\sigma)t}u(t)e^{-j\omega t}dt$$

$$= \text{CTFT}\left\{e^{-(a+\sigma)t}u(t)\right\}.$$

From the known CTFT pairs,

$$e^{-at}u(t), a > 0 \xleftarrow{\text{CTFT}} \frac{1}{a + j\omega}, a > 0$$

Therefore,

$$e^{-(a+\sigma)t}u(t), (a+\sigma) > 0 \xleftarrow{\text{CTFT}} \frac{1}{(a+\sigma) + j\omega}, (a+\sigma) > 0$$

Thus

$$X(s) = \frac{1}{(a+\sigma) + j\omega}, (a+\sigma) > 0$$

$$= \frac{1}{(a+s)}, \sigma > -a$$

$$= \frac{1}{(s+a)}, Re(s) > -a.$$

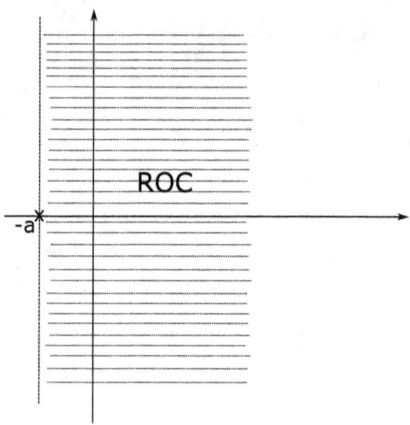

Figure 3.14: ROC: $Re(s) > -a$

The ROC is shown in Figure 3.14.

2. Find the Laplace Transform of $x(t) = -e^{-at}u(-t), a < 0, a$ is a real number.

Solution: Using the analysis equation (Equation (3.36)),

$$X(s) = \int_{-\infty}^{\infty} x(t)e^{-st}dt$$

$$= \int_{-\infty}^{\infty} -e^{-at}u(-t)e^{-st}dt$$

$$= \int_{-\infty}^{\infty} -e^{-at}e^{-\sigma t}u(-t)e^{-j\omega t}dt$$

$$= \int_{-\infty}^{\infty} -e^{-(a+\sigma)t}u(-t)e^{-j\omega t}dt$$

$$= \text{CTFT}\left\{-e^{-(a+\sigma)t}u(-t)\right\}.$$

From the known CTFT pairs,

$$-e^{-at}u(-t), a < 0 \xleftarrow{\text{CTFT}} \frac{1}{a+j\omega}, \ a < 0$$

Therefore,

$$-e^{-(a+\sigma)t}u(-t), (a+\sigma) < 0 \xleftarrow{\text{CTFT}} \frac{1}{(a+\sigma)+j\omega}, \ (a+\sigma) < 0$$

Thus

$$X(s) = \frac{1}{(a+\sigma)+j\omega}, \quad (a+\sigma) < 0$$

$$= \frac{1}{(a+s)}, \quad \sigma < -a$$

$$= \frac{1}{(s+a)}, \quad Re(s) < -a.$$

The ROC is shown in Figure 3.15.

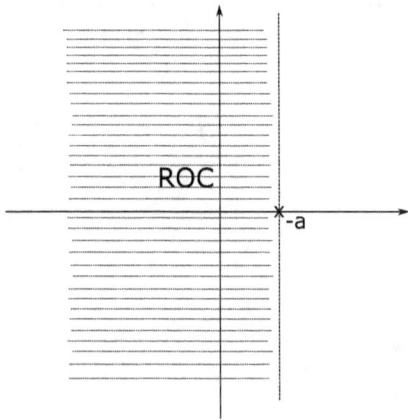

Figure 3.15: ROC: $Re(s) < -a$

3.3.4 Properties of the ROC

1. The ROC of $X(s)$ consists of strips parallel to the $j\omega$ axis in the s−plane.

2. For rational Laplace Transforms, the ROC does not contain any poles.

3. If $x(t)$ is of finite duration and is absolutely integrable, then the ROC is the entire s−plane.

4. If $x(t)$ is right-sided and if the line $Re(s) = \sigma_0$ is in the ROC, then all values of s for which $Re(s) > \sigma_0$ will also be in the ROC.

5. If $x(t)$ is left-sided and if the line $Re(s) = \sigma_0$ is in the ROC, then all values of s for which $Re(s) < \sigma_0$ will also be in the ROC.

6. If $x(t)$ is two-sided and if the line $Re(s) = \sigma_0$ is in the ROC, then the ROC will consist of a strip in the s−plane that includes the line $Re(s) = \sigma_0$.

7. If $X(s)$ is rational then its ROC is bounded by poles or extends to ∞. In addition, no poles of $X(s)$ are contained in the ROC.

8. If $X(s)$ is rational, then if $x(t)$ is right-sided, the ROC is the region in the s−plane to the right of the right-most pole. If $x(t)$ is left-sided, the ROC is the region in the s−plane to the left of the left-most pole.

3.3.5 Common Laplace Transform Pairs

Some commonly used Laplace Transform pairs are given in Table 3.3.

Signal	Laplace Transform	ROC
$\delta(t)$	1	All s
$u(t)$	$\frac{1}{s}$	$Re(s) > 0$
$-u(-t)$	$\frac{1}{s}$	$Re(s) < 0$
$e^{-at}u(t)$	$\frac{1}{s+a}$	$Re(s) > -a$
$-e^{-at}u(-t)$	$\frac{1}{s+a}$	$Re(s) < -a$
$\cos(\omega_0 t)u(t)$	$\frac{s}{s^2+\omega_0^2}$	$Re(s) > 0$
$\sin(\omega_0 t)u(t)$	$\frac{\omega_0}{s^2+\omega_0^2}$	$Re(s) > 0$

Table 3.3: Common Laplace Transform pairs

3.3.6 Properties of Laplace Transform

Consider a signal $x(t)$ with Laplace Transform $X(s)$, that is,

$$x(t) \overset{\text{LT}}{\longleftrightarrow} X(s), \text{ ROC: } R$$

We are interested to know what happens to the Laplace Transform when various signal operations are done in the time (t) domain.

Some of the major properties of the Laplace Transform that are often used when solving numberical problems are discussed next; for a full list of the properties please refer [1].

Linearity

Consider two signals $x(t)$ and $y(t)$ with

$$x(t) \overset{\text{LT}}{\longleftrightarrow} X(s), \text{ ROC: } R_1$$
$$y(t) \overset{\text{LT}}{\longleftrightarrow} Y(s), \text{ ROC: } R_2$$

Then

$$z(t) = Ax(t) + By(t) \overset{\text{LT}}{\longleftrightarrow} Z(s) = AX(s) + BY(s), \qquad (3.37)$$

where $Z(s)$ is the Laplace Transform of $z(t)$. The ROC of $Z(s)$ is: at least $R_1 \cap R_2$. Therefore, *linearity in the time domain* corresponds to *linearity in the s domain*.

Time Shifting

Consider $x(t)$ with

$$x(t) \overset{\text{LT}}{\longleftrightarrow} X(s), \text{ ROC: } R$$

Then

$$x(t - t_0) \overset{\text{LT}}{\longleftrightarrow} e^{-st_0} X(s), \text{ROC: } R \qquad (3.38)$$

Time Scaling

Consider $x(t)$ with

$$x(t) \overset{\text{LT}}{\longleftrightarrow} X(s), \text{ ROC: } R$$

Then

$$x(\alpha t) \overset{\text{LT}}{\longleftrightarrow} \frac{1}{|\alpha|} X(\frac{s}{\alpha}) \qquad (3.39)$$

The ROC is a scaled version of R, that is, s is in the ROC if $\frac{s}{\alpha}$ is in R.

Convolution

Consider $x(t)$ and $y(t)$ with

$$x(t) \overset{\text{LT}}{\longleftrightarrow} X(s), \text{ ROC: } R_1$$
$$y(t) \overset{\text{LT}}{\longleftrightarrow} Y(s), \text{ ROC: } R_2$$

Then

$$z(t) = x(t) * y(t) \overset{\text{LT}}{\longleftrightarrow} Z(s) = X(s)Y(s), \qquad (3.40)$$

where $Z(s)$ is the Laplace Transform of $z(t)$. The ROC of $Z(s)$ is: at least $R_1 \cap R_2$. Therefore, *convolution in the time domain* corresponds to *multiplication in the s domain*.

Conjugation

Consider $x(t)$ with

$$x(t) \overset{\text{LT}}{\longleftrightarrow} X(s), \text{ ROC: } R$$

Then

$$x^*(t) \overset{\text{LT}}{\longleftrightarrow} X^*(s^*), \text{ ROC: } R \qquad (3.41)$$

Shifting in the s Domain

Consider $x(t)$ with

$$x(t) \overset{\text{LT}}{\longleftrightarrow} X(s), \text{ ROC: } R$$

Then

$$e^{s_0 t} x(t) \overset{\text{LT}}{\longleftrightarrow} X(s - s_0). \qquad (3.42)$$

The ROC is a shifted version of R, that is, s is in the ROC if $s - s_0$ is in R.

Differentiation in Time Domain

Consider $x(t)$ with

$$x(t) \overset{\text{LT}}{\longleftrightarrow} X(s), \text{ ROC: } R$$

Then

$$\frac{d}{dt} x(t) \overset{\text{LT}}{\longleftrightarrow} sX(s), \text{ ROC: at least} R \qquad (3.43)$$

Differentiation in s Domain

Consider $x(t)$ with

$$x(t) \overset{\text{LT}}{\longleftrightarrow} X(s), \text{ ROC: } R$$

Then

$$- tx(t) \overset{\text{LT}}{\longleftrightarrow} \frac{d}{ds} X(s), \text{ ROC: } R \qquad (3.44)$$

3.3.7 Solved Examples

1. Find the Laplace Transform of $x(t) = 3e^{-2t}u(t) - 2e^{-t}u(t)$.

Solution: From the known transform

$$e^{-at}u(t) \overset{\text{LT}}{\longleftrightarrow} \frac{1}{s + a}, \quad Re(s) > -a$$

Therefore,

$$e^{-2t}u(t) \xleftrightarrow{\text{LT},a=2} \frac{1}{s+2}, \ Re(s) > -2$$

$$e^{-t}u(t) \xleftrightarrow{\text{LT},a=1} \frac{1}{s+1}, \ Re(s) > -1$$

Thus, using the linearity property,

$$X(s) = 3\frac{1}{s+2} - 2\frac{1}{s+1}, \ Re(s) > -1.$$

The ROC is shown in Figure 3.16 [1].

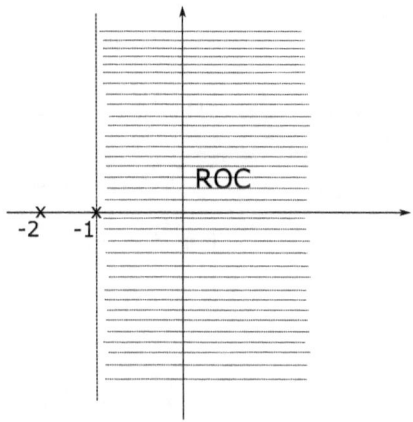

Figure 3.16: ROC: $Re(s) > -1$

2. Find the Laplace Transform of

$$x(t) = \begin{cases} e^{-at}, & 0 < t < T \\ 0, & \text{otherwise} \end{cases}$$

Solution: Using the analysis equation (Equation (3.36)),

---

[1]The common region for $Re(s) > -2$ and $Re(s) > -1$ is $Re(s) > -1$.

$$X(s) = \int_{-\infty}^{\infty} x(t)e^{-st}dt$$

$$= \int_{0}^{T} e^{-at}e^{-st}dt$$

$$= \int_{0}^{T} e^{-(a+s)t}dt$$

$$= \frac{1}{-(a+s)} \left[e^{-(a+s)t} \right]_{0}^{T}$$

$$= \frac{1}{-(a+s)} [e^{-(a+s)T} - 1]$$

$$X(s) = \frac{1}{s+a}[1 - e^{-(s+a)T}].$$

From Property 3 of the ROC (Section 3.3.4), the ROC of $X(s)$ is the entire s−plane.

3. Find the Laplace Transform of $x(t) = -e^{-b|t|}$, b is a real positive number.

Solution: To remove the modulus sign, $x(t)$ can be re-written as:

$$x(t) = \begin{cases} e^{-bt}, & t > 0 \\ e^{bt}, & t < 0 \end{cases}$$

Thus,

$$x(t) = e^{bt}u(-t) + e^{-bt}u(t)$$
$$= A + B$$

To solve for A, from the known Laplace transform pair,

$$-e^{-at}u(-t) \overset{LT}{\longleftrightarrow} \frac{1}{s+a}, \quad Re(s) < -a$$

Setting $a = -b$ and using the linearity property (multiplication by a constant),

$$e^{bt}u(-t) \overset{LT}{\longleftrightarrow} \frac{1}{s-b}, \quad Re(s) < b.$$

To solve for B, from the known Laplace transform pair,

$$e^{-at}u(t) \overset{LT}{\longleftrightarrow} \frac{1}{s+a}, \quad Re(s) > -a$$

Setting $a = b$,

$$e^{-bt}u(t) \overset{\text{LT}}{\longleftrightarrow} \frac{1}{s+b}, \quad Re(s) > -b.$$

Therefore,

$$X(s) = \frac{-1}{s-b} + \frac{1}{s+b}.$$

The ROC is $-b < Re(s) < b$ [2]. This is shown in Figure 3.17.

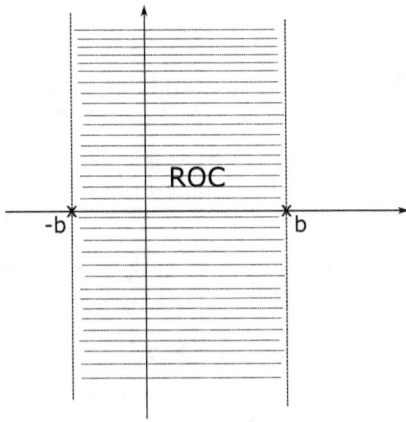

Figure 3.17: ROC: $-b < Re(s) < b$

---

4. Find the Laplace Transform of $x(t) = -e^{-a(t+1)}u(t+1), a > 0, a$ is a real number.

Solution: From the known Laplace transform pair,

$$e^{-at}u(t), a > 0 \overset{\text{LT}}{\longleftrightarrow} \frac{1}{s+a}, \quad Re(s) > -a.$$

Comparing with the time-shifting property (Section 3.3.6) with $t_0 = -1$,

$$e^{-a(t+1)}u(t+1) \overset{\text{LT}}{\longleftrightarrow} e^{-s(-1)}\left(\frac{1}{s+a}\right), \quad Re(s) > -a$$

$$X(s) = e^s\left(\frac{1}{s+a}\right), \quad Re(s) > -a.$$

---

[2]The region common to $Re(s) < b$ and $Re(s) > -b$ is $-b < Re(s) < b$

5.

$$X(s) = \frac{1}{(s+1)(s+2)}, \quad Re(s) > -1$$

Find $x(t)$.

Solution: Using the partial fraction expansion method to find the inverse Laplace Transform,

$$X(s) = \frac{1}{(s+1)(s+2)}$$
$$= \frac{A}{s+1} + \frac{B}{s+2}$$

Solving for A and B, gives $A = 1, B = -1$.

Thus,

$$X(s) = \frac{1}{s+1} - \frac{1}{s+2}, \quad Re(s) > -1$$

From the known Laplace Transform pairs,

$$\frac{1}{s+a}, \quad Re(s) > -a \overset{LT}{\longleftrightarrow} e^{-at}u(t), a > 0$$

If $a = 1$, then

$$\frac{1}{s+1}, \quad Re(s) > -1 \overset{LT}{\longleftrightarrow} e^{-t}u(t).$$

If $a = 2$, then

$$\frac{1}{s+2}, \quad Re(s) > -2 \overset{LT}{\longleftrightarrow} e^{-2t}u(t).$$

Note that the above known transforms are selected such that the common ROC is $Re(s) > -1$. Thus,

$$x(t) = e^{-t}u(t) - e^{-2t}u(t).$$

6.

$$X(s) = \frac{1}{(s+1)(s+2)}, \quad Re(s) > -2$$

Find $x(t)$.

From the previous problem,

$$X(s) = \frac{1}{s+1} - \frac{1}{s+2}, \; Re(s) < -2$$

From the known Laplace Transform pair,

$$\frac{1}{s+a}, \; Re(s) < -a \overset{\text{LT}}{\longleftrightarrow} -e^{-at}u(-t), a > 0$$

If $a = 1$, then

$$\frac{1}{s+1}, \; Re(s) < -1 \overset{\text{LT}}{\longleftrightarrow} -e^{-t}u(-t).$$

If $a = 2$, then

$$\frac{1}{s+2}, \; Re(s) < -2 \overset{\text{LT}}{\longleftrightarrow} -e^{-2t}u(-t).$$

Note that the above known transforms are selected such that the common ROC is $Re(s) < -2$. Thus,

$$x(t) = -e^{-t}u(-t) + e^{-2t}u(-t).$$

3.4 Discrete Time Fourier Series (DTFS)

In DTFS, we consider a discrete time periodic signal $x[n]$ with fundamental frequency Ω_0 and fundamental period N.

The basis signals chosen are complex exponentials (refer Section 1.5) of the form $e^{jk\Omega_0 n}$, where Ω_0 is the fundamental frequency and $k = 0, \pm 1, \pm 2, \ldots$. The basis signals are thus harmonically related with the fundamental frequency Ω_0 (meaning, $k\Omega_0$ is the kth harmonic of Ω_0.). Note that the basis signals are also periodic signals with fundamental frequency Ω_0.

The time domain signal $x[n]$ can be written as a linear combination of the above basis signals as:

$$x[n] = \sum_{k=<N>} a_k e^{jk\Omega_0 n}, \tag{3.45}$$

where a_k are complex constants and are called the Fourier Series coefficients or spectral coefficients of $x[n]$.

Equation (3.45) is the Fourier Series representation of the periodic signal $x[n]$. Another way of looking at Equation (3.45) is, if we know Ω_0, k (integers from $-\infty$ to $+\infty$) and a_k (for each value of k), meaning the right hand side of Equation (3.45)), then we can compute the time domain signal $x[n]$. In this sense we are transforming

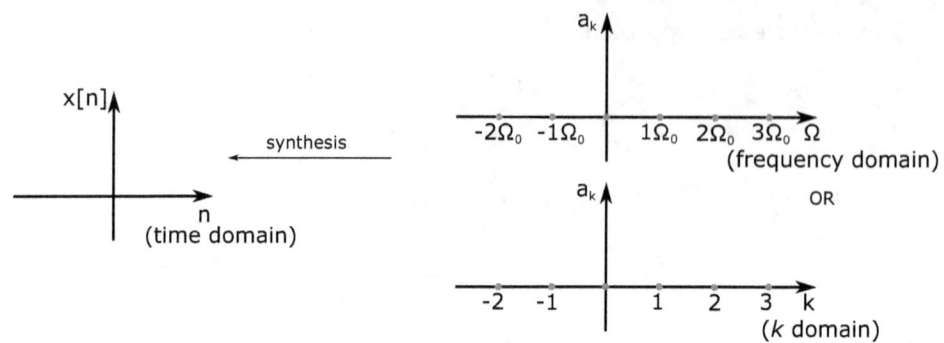

Figure 3.18: DTFS synthesis equation

the signal from the Ω (frequency domain, or equivalently k domain) to the time domain. This is shown in Figure 3.18.

Note that k takes on only N values. This is because there are only N distinct signals in the set $e^{jk\Omega_0 n}$. That is,

$$e^{j0\Omega_0 n}(\text{when } k = 0) = e^{jN\Omega_0 n}(\text{when } k = N) \; \forall n.$$

Similarly

$$e^{j1\Omega_0 n}(\text{when } k = 1) = e^{j(N+1)\Omega_0 n}(\text{when } k = N+1) \; \forall n.$$

The expression for a_k is given by:

$$a_k = \frac{1}{N} \sum_{n=<N>} x[n]e^{-jk\omega_0 n}. \tag{3.46}$$

Another way looking at Equation (3.46) is, if we know Ω_0, k (integers from $-\infty$ to $+\infty$) and $x[n]$, meaning the right hand side of Equation (3.46)), then we can compute a_k (for each value of k). In this sense we are transforming the signal from the time domain to the Ω (frequency domain, or equivalently k domain). This is shown in Figure 3.19.

As mentioned above, there are only N distinct values for k. *Therefore a_k is periodic in N.* Note that this is not true in CTFS.

To summarize, the two equations for DTFS (called the analysis and synthesis equations) are given by:

112

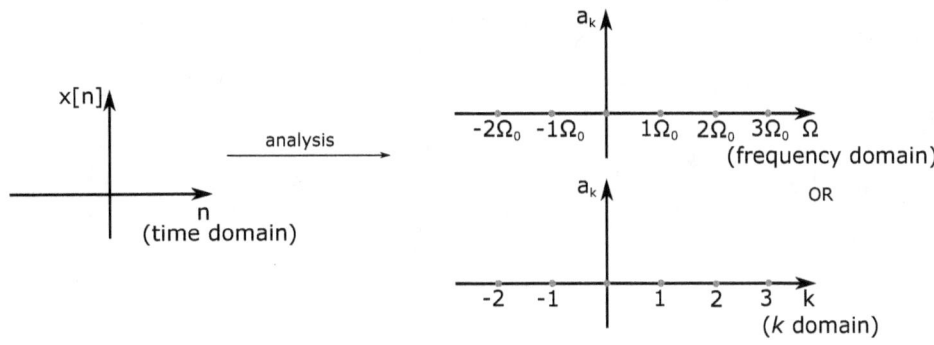

Figure 3.19: DTFS analysis equation

Synthesis equation:

$$x[n] = \sum_{k=<N>} a_k e^{jk\Omega_0 n} \tag{3.47}$$

Analysis equation:

$$a_k = \frac{1}{N} \sum_{n=<N>} x[n] e^{-jk\Omega_0 n}, \text{ periodic in } N \tag{3.48}$$

Transform represented by:

$$x[n] \xleftrightarrow{\text{DTFS}} a_k$$

Just like in CTFS, there are two methods to solve numerical problems related to finding the DTFS coefficients a_k. The first method, a comparision (to Equation (3.47)) method is used for problems where the time-domain signal is trigonometric function. For other types of signals, the second method of directly applying Equation (3.48) is used. Both these methods are demonstrated in the examples below.

3.4.1 Solved Examples

1. Find the DTFS coefficients of $x[n] = \sin(\Omega_0 n)$.

113

Solution: Using the comparision method

$$x[n] = \sin(\Omega_0 n)$$

$$= \frac{e^{j\Omega_0 n} - e^{-j\Omega_0 n}}{2j}$$

$$= \frac{1}{2j} e^{j\Omega_0 n} - \frac{1}{2j} e^{-j\Omega_0 n}$$

Comparing with Equation (3.47), $a_1 = \frac{1}{2j}, a_{-1} = \frac{-1}{2j}$. Importantly, the coefficients are periodic in N.

2. For the DT periodic square wave shown in Figure 3.20, find the DTFS coefficients.

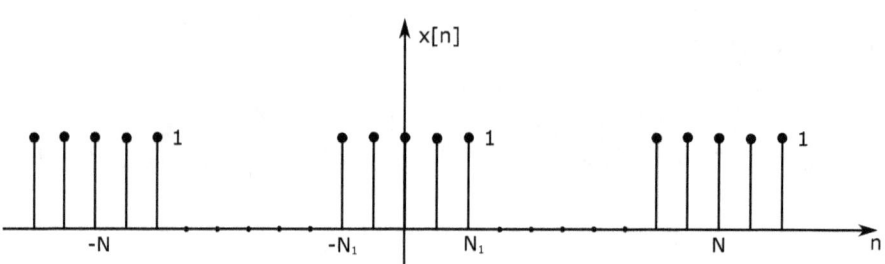

Figure 3.20: A DT periodic square wave

Solution: Using the analysis equation, we will first solve for the case when $k = 0$, then solve for when $k \neq 0$.

When $k = 0$,

$$a_k = \frac{1}{N} \sum_{n=<N>} x[n] e^{-jk\Omega_0 n}$$

$$a_0 = \frac{1}{N} \sum_{n=-N}^{N} x[n]$$

$$= \frac{1}{N} \sum_{n=-N_1}^{N_1} x[n]$$

$$= \frac{1}{N} \sum_{n=-N_1}^{N_1} 1$$

$$= \frac{1}{N} (2N_1 + 1)$$

$$= \frac{2N_1 + 1}{N}.$$

When $k \neq 0$,

$$a_k = \frac{1}{N} \sum_{n=<N>} x[n] e^{-jk\Omega_0 n}$$

$$= \frac{1}{N} \sum_{n=-N_1}^{N_1} (1) e^{-jk\Omega_0 n}$$

Using a change in variables, let $m = n + N_1$ or $n = m - N_1$. Therefore when $n = -N_1, m = 0$ and when $n = N_1, m = 2N_1$. Thus,

$$a_k = \frac{1}{N} \sum_{m=0}^{2N_1} e^{-jk\Omega_0(m-N_1)}$$

$$= \frac{1}{N} \sum_{m=0}^{2N_1} e^{-jk\Omega_0 m} e^{+jk\Omega_0 N_1}$$

$$= \frac{e^{jk\Omega_0 N_1}}{N} \sum_{m=0}^{2N_1} e^{-jk\Omega_0 m}.$$

Using the following geometric series formula,

$$\sum_{n=0}^{N-1} \alpha^n = \frac{1 - \alpha^{(N-1)}}{1 - \alpha}, \ \alpha \neq 1$$

115

$$a_k = \frac{e^{jk\Omega_0 N_1}}{N} \left[\frac{1 - (e^{-jk\Omega_0})^{(2N_1+1)}}{1 - e^{jk\Omega_0}} \right].$$

This can be further simplified to

$$a_k = \frac{e^{jk\Omega_0 N_1}}{N} \frac{\sin\left[k\Omega_0(N_1 + 1/2)\right]}{\sin(k\Omega_0/2)}.$$

---

3. Determine the DTFS representation of the following sequences: $x[n] = \cos(\pi/3n) + \sin(\pi/4n)$.

Solution: First, we need to check if $x[n]$ is a periodic signal. Since $x[n]$ is a sum of two sinusoidal signals, we will use the approach seen earlier in Section 1.10.

For $\cos(\pi/3n)$,

$$\cos(\pi/3n) \Rightarrow \Omega = \frac{\pi}{3}$$

$$N = m\frac{2\pi}{\Omega} = m\frac{2\pi}{\frac{\pi}{3}} = 6m.$$

For $m = 1$, the fundamental period $N_1 = 6$ (an integer number). Thus $\cos(\pi/3n)$ is periodic.

For $\sin(\pi/4n)$,

$$\sin(\pi/4n) \Rightarrow \Omega = \frac{\pi}{4}$$

$$N = m\frac{2\pi}{\Omega} = m\frac{2\pi}{\frac{\pi}{4}} = 8m.$$

For $m = 1$, the fundamental period $N_2 = 8$ (an integer number). Thus $\sin(\pi/4n)$ is periodic.

Therefore,

$$\frac{N_1}{N_2} = \frac{6}{8}.$$

Since $\frac{N_1}{N_2}$ is a rational number, the signal $x[n]$ is periodic. The periodicity is given by LCM(6,8) = 24 and the fundamental frequency $\Omega_0 = \frac{2\pi}{24}$.

Since $x[n]$ is a trigonometric function, we will be using the comparision method.

Expanding $x[n]$ in terms of Ω_0,

$$
\begin{aligned}
x[n] &= \cos(\pi/3n) + \sin(\pi/4n) \\
&= \cos(4\frac{2\pi}{24}n) + \sin(3\frac{2\pi}{24}n) \\
&= \cos(4\Omega_0 n) + \sin(3\Omega_0 n) \\
&= \frac{e^{j4\Omega_0 n} + e^{-j4\Omega_0 n}}{2} + \frac{e^{j3\Omega_0 n} - e^{-j3\Omega_0 n}}{2j} \\
&= \frac{1}{2}e^{j4\Omega_0 n} + \frac{1}{2}e^{-j4\Omega_0 n} + \frac{1}{2j}e^{j3\Omega_0 n} - \frac{1}{2j}e^{-j3\Omega_0 n}
\end{aligned}
$$

Comparing with Equation (3.47), the DTFS coefficients are: $a_4 = \frac{1}{2}$, $a_{-4} = \frac{1}{2}$, $a_3 = \frac{1}{2j}$, $a_{-3} = \frac{1}{2j}$. Note that the DTFS coefficients are periodic in 24.

---

4. Consider a sequence

$$
x[n] = \sum_{k=-\infty}^{\infty} \delta[n - 4k]
$$

Find the Fourier Series coefficients of x[n].

Solution: A plot of $x[n]$ is shown in Figure 3.21. Thus we see that $x[n]$ is periodic with $N = 4$. $\Omega_0 = \frac{2\pi}{N} = \frac{\pi}{2}$ (for $m = 1$).

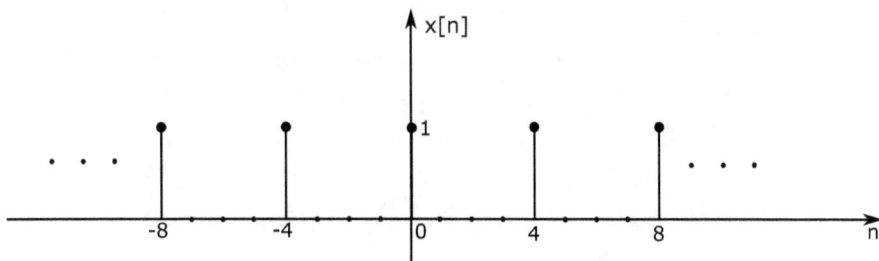

Figure 3.21: Plot of $x[n]$

117

Applying the analysis equation:

$$a_k = \frac{1}{N} \sum_{n=<N>} x[n] e^{-jk\Omega_0 n}$$

$$= \frac{1}{4} \sum_{n=0}^{3} x[n] e^{-jk\frac{\pi}{2}n}$$

$$= \frac{1}{4}[1 + 0 + 0 + 0] \quad (\text{since } x[n] = 0 \text{ for } n = 1, 2, 3)$$

$$= \frac{1}{4}.$$

Note that a_k is periodic in $N = 4$.

3.4.2 Properties of the DTFS

Consider a periodic signal $x[n]$ with Fourier Series coefficients a_k, that is,

$$x[n] \xleftrightarrow{\text{DTFS}} a_k.$$

We are interested to know what happens to the Fourier Series coefficients or what happens in the k domain when various signal operations are done in the time (n) domain. Note that all DTFS coefficients are periodic in N.

Some of the major properties of the DTFS that are often used when solving numerical problems are discussed next; for a full list of the properties please refer [1].

Linearity

If $x[n]$ and $y[n]$ are periodic signals with period N and

$$x[n] \xleftrightarrow{\text{DTFS}} a_k \quad (\text{periodic in } N)$$
$$y[n] \xleftrightarrow{\text{DTFS}} b_k \quad (\text{periodic in } N)$$

Then

$$z[n] = Ax[n] + By[n] \xleftrightarrow{\text{DTFS}} c_k = Aa_k + Bb_k, \tag{3.49}$$

where c_k are the Fourier Series coefficients of $z[n]$.

Therefore, *linearity in the time domain* corresponds to *linearity in the k domain.*

Time Shifting

If $x[n]$ is a periodic signal with period N and

$$x[n] \xleftrightarrow{\text{DTFS}} a_k \quad (\text{periodic in } N)$$

Then

$$x[n - n_0] \xleftrightarrow{\text{DTFS}} b_k = e^{-jk\frac{2\pi}{N}n_0} a_k \tag{3.50}$$

where b_k are the Fourier Series coefficients of $x[n - n_0]$. Note that $x[n - n_0]$ is also periodic in N.

Therefore, *shifting in the time domain* corresponds to *phase shift in the k domain*. Note that the magnitude spectrum of a_k and b_k are the same.

Time Reversal

If $x[n]$ is a periodic signal with period N and

$$x[n] \xleftrightarrow{\text{DTFS}} a_k \quad (\text{periodic in } N)$$

Then

$$x[-n] \xleftrightarrow{\text{DTFS}} b_k = a_{-k} \tag{3.51}$$

where b_k are the Fourier Series coefficients of $x[-n]$. Note that $x[-n]$ is also periodic in N.

Therefore, *reversal in the time domain* corresponds to *reversal in the k domain*.

Time Scaling

If $x[n]$ is a periodic signal with period N and

$$x[n] \xleftrightarrow{\text{DTFS}} a_k \quad (\text{periodic in } N)$$

Then

$$x_{(m)}[n] \xleftrightarrow{\text{DTFS}} b_k = \frac{1}{m} a_k \quad (\text{periodic in } mN) \tag{3.52}$$

where b_k are the Fourier Series coefficients of $x_{(m)}[n]$. Note that $x_{(m)}[n]$ is defined as:

$$x_{(m)}[n] = \begin{cases} x[n/m] & \text{, if } n \text{ is a multiple of } m \\ \\ 0 & \text{, if } n \text{ is not a multiple of } m \end{cases}$$

and is periodic in mN.

Multiplication

If $x[n]$ and $y[n]$ are periodic signals with period N and

$$x[n] \xleftrightarrow{\text{DTFS}} a_k \quad (\text{periodic in } N)$$
$$y[n] \xleftrightarrow{\text{DTFS}} b_k \quad (\text{periodic in } N)$$

Then

$$z[n] = x[n]y[n] \xleftrightarrow{\text{DTFS}} c_k = \sum_{l=<N>} a_l b_{k-l} \qquad (3.53)$$

where c_k are the Fourier Series coefficients of $z[n]$.

Therefore, *multiplication in the time domain* corresponds to *convolution in the k domain*.

Conjugation

If $x[n]$ is a periodic signal with period N and

$$x[n] \xleftrightarrow{\text{DTFS}} a_k \quad (\text{periodic in } N)$$

Then

$$x^*[n] \xleftrightarrow{\text{DTFS}} b_k = a^*_{-k} \qquad (3.54)$$

where b_k are the Fourier Series coefficients of $x^*[n]$. Note that $x^*[n]$ is periodic in N.

Frequency Shift

If $x[n]$ is a periodic signal with period N and

$$x[n] \xleftrightarrow{\text{DTFS}} a_k \quad (\text{periodic in } N)$$

Then

$$e^{jM\frac{2\pi}{N}n}x[n] \xleftrightarrow{\text{DTFS}} b_k = a_{k-M} \qquad (3.55)$$

where b_k are the Fourier Series coefficients of $e^{jM\frac{2\pi}{N}n}x[n]$.

Therefore, *phase shift in the time domain* corresponds to *frequency shift in the k domain*.

Parseval's Theorem

If $x[n]$ is a periodic signal with period N and

$$x[n] \xleftrightarrow{\text{DTFS}} a_k \quad (\text{periodic in } N)$$

Then

$$\frac{1}{N} \sum_{n=<N>} |x[n]|^2 = \sum_{k=<N>} |a_k|^2. \tag{3.56}$$

3.5 Discrete Time Fourier Transform (DTFT)

In DTFT, we consider a discrete time aperiodic signal $x[n]$. We are going to represent $x[n]$ as a linear combination of basis signals. The signals chosen are complex exponentials of the form $e^{j\omega n}$.

The time domain signal $x[n]$ can be written as a linear combination of the above basis signals as:

$$x[n] = \frac{1}{2\pi} \int_{<2\pi>} X(e^{j\omega}) e^{j\omega n} \, d\omega, \tag{3.57}$$

where $X(e^{j\omega})$ is called the Fourier Transform or spectrum of $x[n]$.

Another way of looking at Equation (3.57) is, if we know ω and $X(e^{j\omega})$, meaning the right hand side of Equation (3.57)), then we can compute the time domain signal $x[n]$. In this sense we are transforming the signal from the ω (frequency domain) to the time domain. This is shown in Figure 3.22.

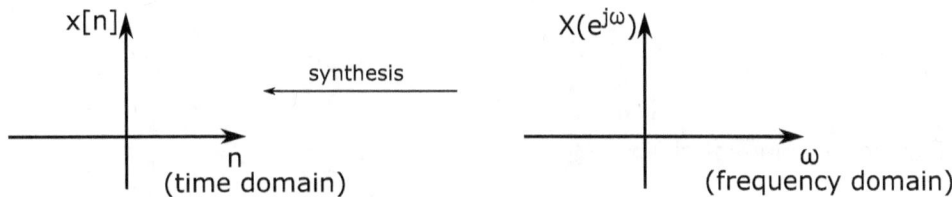

Figure 3.22: DTFT synthesis equation

Note that $e^{j\omega}$ is periodic in 2π. That is,

$$e^{j(\omega+2\pi)n} = e^{j\omega n} e^{j2\pi n} = e^{j\omega n} (\text{since } e^{j2\pi n} = 1 \; \forall n).$$

Thus $X(e^{j\omega})$ is periodic in 2π.

The expression for $X(e^{j\omega})$ is given by

$$X(e^{j\omega}) = \sum_{n=-\infty}^{\infty} x[n] e^{-j\omega n}. \tag{3.58}$$

Another way looking at Equation (3.58) is, if we know ω and $x[n]$, meaning the right hand side of Equation (3.58)), then we can compute $X(e^{j\omega})$. In this sense we are transforming the signal from the n (time) domain to the ω (frequency) domain. This is shown in Figure 3.23. As mentioned, $X(e^{j\omega})$ is periodic in 2π.

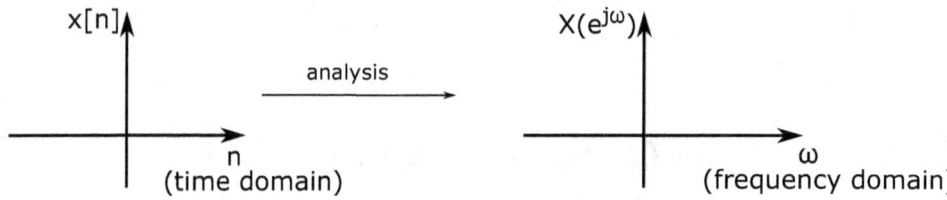

Figure 3.23: DTFT analysis equation

To summarize, the two equations for DTFT (called the analysis and synthesis equations) are given by:

Synthesis equation:

$$x[n] = \frac{1}{2\pi} \int_{<2\pi>} X(e^{j\omega}) \, e^{j\omega n} \, d\omega \qquad (3.59)$$

Analysis equation:

$$X(e^{j\omega}) = \sum_{n=-\infty}^{\infty} x[n]e^{-j\omega n}, \text{ periodic in } 2\pi \qquad (3.60)$$

Transform represented by:

$$x[n] \xleftrightarrow{\text{DTFT}} X(e^{j\omega})$$

3.5.1 Solved Examples

1. Find the DTFT of $x[n] = a^n u[n], |a| < 1$.

Solution: Using Equation (3.48),

$$X(e^{j\omega}) = \sum_{n=-\infty}^{\infty} x[n]e^{-j\omega n}$$

$$= \sum_{n=-\infty}^{\infty} a^n u[n]e^{-j\omega n}$$

$$= \sum_{n=0}^{\infty} a^n e^{-j\omega n} \text{ (using the definition of } u[n])$$

$$= \sum_{n=0}^{\infty} (ae^{-j\omega})^n$$

This is the sum of infinite terms of a geometric series, which is given by:

$$1 + \alpha + \alpha^2 + \alpha^3 + \ldots = \frac{1}{1-\alpha}, \ |\alpha| < 1.$$

Therefore,

$$X(e^{j\omega}) = \frac{1}{1 - ae^{-j\omega}}.$$

2. Find the DTFT of $x[n] = a^{|n|}, |a| < 1$.

Solution: Using Equation (3.48),

$$X(e^{j\omega}) = \sum_{n=-\infty}^{\infty} x[n]e^{-j\omega n}$$

$$= \sum_{n=-\infty}^{\infty} a^{|n|} e^{-j\omega n}$$

$$= \sum_{n=-\infty}^{-1} a^{-n} e^{-j\omega n} + \sum_{n=0}^{\infty} a^n e^{-j\omega n}$$

$$= A + B$$

Calculating A and B,

$$A = \sum_{n=-\infty}^{-1} a^{-n} e^{-j\omega n}$$

Using a change of variables, let $m = -n$; when $n = -\infty, m = \infty$ and $n = -1, m = 1$.

$$
\begin{aligned}
A &= \sum_{m=\infty}^{1} a^m e^{j\omega m} \\
&= \sum_{m=1}^{\infty} a^m e^{j\omega m} \\
&= \sum_{m=1}^{\infty} (ae^{j\omega})^m \\
&= \frac{ae^{j\omega}}{1 - ae^{j\omega}} \quad \text{(using the geometric series formula)}
\end{aligned}
$$

$$
\begin{aligned}
B &= \sum_{n=0}^{\infty} (ae^{-j\omega})^n \\
&= \frac{1}{1 - ae^{j\omega}} \quad \text{(using the geometric series formula)}
\end{aligned}
$$

Therefore

$$
\begin{aligned}
X(e^{j\omega}) &= \frac{ae^{j\omega}}{1 - ae^{j\omega}} + \frac{1}{1 - ae^{j\omega}} \\
&= \frac{ae^{j\omega}(1 - ae^{-j\omega}) + (1 - ae^{j\omega})}{(1 - ae^{j\omega})(1 - ae^{-j\omega})} \\
&= \frac{1 - a^2}{(1 - ae^{j\omega})(1 - ae^{-j\omega})} \\
X(e^{j\omega}) &= \frac{1 - a^2}{1 - 2a\cos(\omega) + a^2}.
\end{aligned}
$$

3.5.2 Common DTFT Pairs

Some commonly used DTFT pairs are given in Table 3.4. For a more elaborate list of common DTFT pairs, please refer [1].

Signal	Fourier Transform
$x[n] = 1$	$2\pi \sum_{l=-\infty}^{+\infty} \delta(\omega - 2\pi l)$
$\delta[n]$	1
$u[n]$	$\frac{1}{1-e^{-j\omega}} + \sum_{k=-\infty}^{+\infty} \pi\delta(\omega - 2\pi k)$
$\cos(\omega_0 n)$	$\pi \sum_{l=-\infty}^{+\infty} [\delta(\omega - \omega_0 - 2\pi l) + \delta(\omega + \omega_0 - 2\pi l)]$
$\sin(\omega_0 n)$	$\frac{\pi}{j} \sum_{l=-\infty}^{+\infty} [\delta(\omega - \omega_0 - 2\pi l) - \delta(\omega + \omega_0 - 2\pi l)]$
$a^n u[n], |a| < 1$	$\frac{1}{1-ae^{-j\omega}}$
$(n+1)a^n u[n], |a| < 1$	$\frac{1}{(1-ae^{-j\omega})^2}$

Table 3.4: Common DTFT pairs

3.5.3 Properties of the DTFT

Consider a signal $x[n]$ with Fourier coefficients $X(e^{j\omega})$, that is,

$$x[n] \xleftrightarrow{\text{DTFT}} X(e^{j\omega}).$$

We are interested to know what happens to the Fourier coefficients or what happens in the ω domain when various signal operations are done in the time (n) domain. Note that all DTFT coefficients are periodic in 2π.

Some of the major properties of the DTFT that are often used when solving numerical problems are discussed next; for a full list of the properties please refer [1].

Linearity

Consider two signals $x[n]$ and $y[n]$ with

$$x[n] \xleftrightarrow{\text{DTFT}} X(e^{j\omega}) \text{ (periodic in } 2\pi)$$
$$y[n] \xleftrightarrow{\text{DTFT}} Y(e^{j\omega}) \text{ (periodic in } 2\pi)$$

Then

$$z[n] = Ax[n] + By[n] \xleftrightarrow{\text{DTFT}} Z(e^{j\omega}) = AX(e^{j\omega}) + BY(e^{j\omega}), \qquad (3.61)$$

where $Z(e^{j\omega})$ are the Fourier coefficients of $z[n]$.

Therefore, *linearity in the time domain* corresponds to *linearity in the ω domain*.

Time Shifting

Consider $x[n]$ with

$$x[n] \xleftrightarrow{\text{DTFT}} X(e^{j\omega}) \text{ (periodic in } 2\pi)$$

Then

$$x[n - n_0] \xleftrightarrow{\text{DTFT}} e^{-j\omega n_0} X(e^{j\omega}) \tag{3.62}$$

Therefore, *shifting in the time domain* corresponds to *phase shift in the ω domain*.

Time Reversal

Consider $x[n]$ with

$$x[n] \xleftrightarrow{\text{DTFT}} X(e^{j\omega}) \text{ (periodic in } 2\pi)$$

Then

$$x[-n] \xleftrightarrow{\text{DTFT}} X(e^{-j\omega}) \tag{3.63}$$

Therefore, *reversal in the time domain* corresponds to *reversal in the ω domain*.

Time Expansion

Consider $x[n]$ with

$$x[n] \xleftrightarrow{\text{DTFT}} X(e^{j\omega}) \text{ (periodic in } 2\pi)$$

Then

$$x_{(k)}[n] \xleftrightarrow{\text{DTFT}} X(e^{jk\omega}) \tag{3.64}$$

Note that $x_{(k)}[n]$ is defined as:

$$x_{(k)}[n] = \begin{cases} x[n/k] & \text{, if } n \text{ is a multiple of } k \\ \\ 0 & \text{, if } n \text{ is not a multiple of } k \end{cases}$$

Multiplication

Consider $x[n]$ and $y[n]$ with

$$x[n] \xleftrightarrow{\text{DTFT}} X(e^{j\omega}) \text{ (periodic in } 2\pi)$$
$$y[n] \xleftrightarrow{\text{DTFT}} Y(e^{j\omega}) \text{ (periodic in } 2\pi)$$

Then

$$z[n] = x[n]y[n] \xleftrightarrow{\text{DTFT}} \frac{1}{2\pi} \int_{2\pi} X(e^{j\theta})Y(e^{j(\omega-\theta)})d\theta \qquad (3.65)$$

Therefore, *multiplication in the time domain* corresponds to *convolution in the ω domain*.

Convolution

Consider $x[n]$ and $y[n]$ with

$$x[n] \xleftrightarrow{\text{DTFT}} X(e^{j\omega}) \text{ (periodic in } 2\pi)$$
$$y[n] \xleftrightarrow{\text{DTFT}} Y(e^{j\omega}) \text{ (periodic in } 2\pi)$$

Then

$$x[n] * y[n] \xleftrightarrow{\text{DTFT}} X(e^{j\omega})Y(e^{j\omega}) \qquad (3.66)$$

Therefore, *convolution in the time domain* corresponds to *multiplication in the ω domain*.

Conjugation

Consider $x[n]$ with

$$x[n] \xleftrightarrow{\text{DTFT}} X(e^{j\omega}) \text{ (periodic in } 2\pi)$$

Then

$$x^*[n] \xleftrightarrow{\text{DTFT}} X^*(e^{-j\omega}) \qquad (3.67)$$

Frequency Shift

Consider $x[n]$ with

$$x[n] \xleftrightarrow{\text{DTFT}} X(e^{j\omega}) \text{ (periodic in } 2\pi)$$

Then

$$e^{j\omega_0 n}x[n] \xleftrightarrow{\text{DTFT}} X(e^{j(\omega-\omega_0)}) \qquad (3.68)$$

Therefore, *phase shift in the time domain* corresponds to *frequency shift in the ω domain*.

Parseval's Theorem

Consider $x[n]$ with

$$x[n] \xleftrightarrow{\text{DTFT}} X(e^{j\omega}) \quad (\text{periodic in } 2\pi)$$

Then

$$\sum_{n=-\infty}^{+\infty} |x[n]|^2 = \frac{1}{2\pi} \int_{2\pi} |X(e^{j\omega})|^2 d\omega \qquad (3.69)$$

3.5.4 Solved Examples

1. Consider a LTI system with $h[n] = \alpha^n u[n], |\alpha| < 1$. For an input $x[n] = \beta^n u[n], |\beta| < 1, \alpha \neq \beta$, find $y[n]$.

Solution: The system block diagram is shown in Figure 3.24.

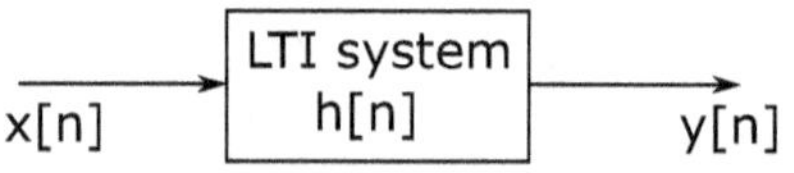

Figure 3.24: A DT LTI system

Given:

$$h[n] = \alpha^n u[n], \ |\alpha| < 1$$
$$x[n] = \beta^n u[n], \ |\beta| < 1, \alpha \neq \beta$$
$$y[n] = ?$$

Since the system is LTI,

$$y[n] = x[n] * h[n]$$
$$Y(e^{j\omega}) = X(e^{j\omega})H(e^{j\omega})$$

From the known DTFT transforms,

$$x[n] = \beta^n u[n], \ |\beta| < 1 \xleftrightarrow{\text{DTFT}} X(e^{j\omega}) = \frac{1}{1 - \beta e^{-j\omega}}$$
$$h[n] = \alpha^n u[n], \ |\alpha| < 1 \xleftrightarrow{\text{DTFT}} H(e^{j\omega}) = \frac{1}{1 - \alpha e^{-j\omega}}$$

Therefore

$$Y(e^{j\omega}) = \frac{1}{(1 - \alpha e^{-j\omega})(1 - \beta e^{-j\omega})}$$

$$= \frac{A}{1 - \alpha e^{-j\omega}} + \frac{B}{1 - \beta e^{-j\omega}}.$$

Solving for A and B gives $A = \frac{\alpha}{\alpha - \beta}, B = \frac{-\beta}{\alpha - \beta}$.

Therefore

$$Y(e^{j\omega}) = \left(\frac{\alpha}{\alpha - \beta}\right) \frac{1}{1 - \alpha e^{-j\omega}} - \left(\frac{\beta}{\alpha - \beta}\right) \frac{1}{1 - \beta e^{-j\omega}}.$$

From the known DTFT transforms,

$$y[n] = \left(\frac{\alpha}{\alpha - \beta}\right) \alpha^n u[n] - \left(\frac{\beta}{\alpha - \beta}\right) \beta^n u[n].$$

2. Find the DTFT of $x[n] = n(1/2)^n u[n]$.

Solution: From the known DTFT transform,

$$y[n] = \alpha^n u[n], \; |\alpha| < 1 \xleftrightarrow{\text{DTFT}} Y(e^{j\omega}) = \frac{1}{1 - \alpha e^{-j\omega}}, \; \alpha = \frac{1}{2}$$

Using the differentiation in frequency property,

$$x[n] = ny[n] \xleftrightarrow{\text{DTFT}} X(e^{j\omega}) = j\frac{d}{d\omega}Y(e^{j\omega}).$$

Therefore

$$X(e^{j\omega}) = j\frac{d}{d\omega}\left(\frac{1}{1 - \frac{1}{2}e^{-j\omega}}\right)$$

$$= j\frac{d}{d\omega}\left(1 - \frac{1}{2}e^{-j\omega}\right)^{-1}$$

$$= j(-1)\left(1 - \frac{1}{2}e^{-j\omega}\right)^{-2}\left(\frac{-1}{2}\right)e^{-j\omega}(-j)$$

$$X(e^{j\omega}) = \frac{\frac{1}{2}e^{-j\omega}}{\left(1 - \frac{1}{2}e^{-j\omega}\right)^2}.$$

3. Find the DTFT of $x[n] = u[n+1] - u[n-2]$.

Solution: From the known DTFT transform,

$$y[n] = u[n] \xleftrightarrow{\text{DTFT}} Y(e^{j\omega}) = \frac{1}{1 - e^{-j\omega}}$$

Applying the time shifting property,

$$u[n - (-1)] \xleftrightarrow{\text{DTFT}} \frac{e^{-j\omega(-1)}}{1 - e^{-j\omega}}$$

$$u[n - 2] \xleftrightarrow{\text{DTFT}} \frac{e^{-j2\omega}}{1 - e^{-j\omega}}$$

Applying the linearity property,

$$u[n+1] - u[n-2] \xleftrightarrow{\text{DTFT}} \frac{e^{j\omega}}{1 - e^{-j\omega}} - \frac{e^{-j2\omega}}{1 - e^{-j\omega}}$$

$$x[n] \xleftrightarrow{\text{DTFT}} \frac{e^{j\omega} - e^{-j2\omega}}{1 - e^{-j\omega}}.$$

3.6 Z Transform

Consider a discrete time signal $x[n]$. We are going to represent $x[n]$ as a linear combination of basis signals. The basis signals chosen are complex exponentials (refer Section 1.5) of the form Cz^n, where $C, z = re^{j\omega}$ are complex numbers.

The time domain signal $x[n]$ can be written as a linear combination of the above basis signals as:

$$x[n] = \frac{1}{2\pi} \int_{2\pi} X(re^{j\omega})(re^{j\omega})^n d\omega \tag{3.70}$$

In the complex domain, the above equation can be written as:

$$x[n] = \frac{1}{2\pi j} \oint X(z) z^{n-1} dz \tag{3.71}$$

where the symbol $\oint$ denotes integration around a counterclockwise closed circular contour centred at the origin and with radius r. Note that $X(z)$ is called the Z Transform of $x[n]$.

Another way of looking at Equation (3.71) is, if we know z and $X(z)$, meaning the right hand side of Equation (3.71)), then we can compute the time domain signal $x[n]$. In this sense we are transforming the signal from the complex z domain to the time domain. This is shown in Figure 3.25. Note that for the numerical problems in this book, both Equations (3.70) and (3.71) are not used. Instead, the partial-fraction expansion method is used (as done for Laplace Transform).

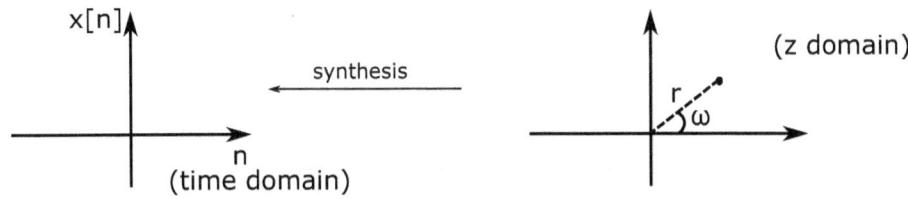

Figure 3.25: Z Transform synthesis equation

The expression for $X(z)$ is given by:

$$X(z) = \sum_{n=-\infty}^{\infty} x[n]z^{-n} = \sum_{n=-\infty}^{\infty} x[n](re^{j\omega})^{-n} \qquad (3.72)$$

Another way looking at Equation (3.72) is, if we know z and $x[n]$, meaning the right hand side of Equation (3.72)), then we can compute $X(z)$ (for each value of z). In this sense we are transforming the signal from the time domain to the complex z domain. This is shown in Figure 3.26. Equation (3.72) is called the Bilateral Z Transform of $x[n]$.

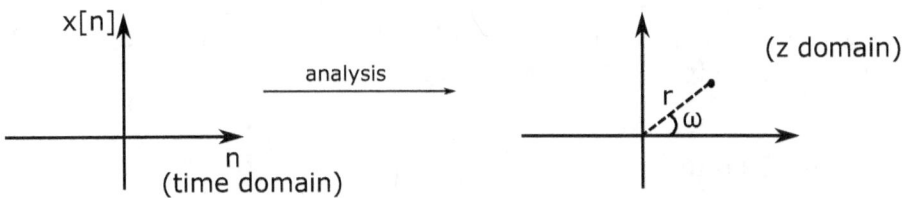

Figure 3.26: Z Transform analysis equation

To summarize, the two equations for the Bilateral Z Transform (called the analysis and synthesis equations) are given by:

131

Synthesis equation:

$$x[n] = \frac{1}{2\pi j} \int X(z)z^{n-1}dz \tag{3.73}$$

Analysis equation:

$$X(z) = \sum_{n=-\infty}^{\infty} x[n]z^{-n} \tag{3.74}$$

Transform represented by:

$$x[n] \overset{Z}{\longleftrightarrow} X(z)$$

3.6.1 DTFT as a Special Case of the Z Transform

From Equation (3.74),

$$X(z) = \sum_{n=-\infty}^{\infty} x[n](re^{j\omega})^{-n}$$

$$= \sum_{n=-\infty}^{\infty} x[n]r^{-n}e^{-j\omega n}$$

$$= \text{DTFT}\left\{x[n]r^{-n}\right\}$$

Thus the Z transform of $x[n]$ is the same as the DTFT of $x[n]r^{-n}$. If $r = 1$ (unit circle), that is $z = e^{j\omega}$, then, $X(z) = \text{DTFT}\{x[n]\}$, that is the Z Transform of $x[n]$ is the same as the DTFT of $x[n]$.

3.6.2 Region of Convergence (ROC)

The range of values of z for which Equation (3.74) converges is called the ROC of the Z Transform. Note that z is a complex number.

3.6.3 Solved Examples

1. $x[n] = a^n u[n]$. Find the Z transform.

Solution: We will solve this problem by directly applying the analysis equation (3.74).

$$X(z) = \sum_{n=-\infty}^{\infty} x[n]z^{-n}$$

$$= \sum_{n=-\infty}^{\infty} a^n u[n]z^{-n}$$

$$= \sum_{n=0}^{\infty} (az^{-1})^n \quad \text{(using the definition of } u[n])$$

This is the sum of infinite terms of a geometric series, which is given by:

$$1 + \alpha + \alpha^2 + \alpha^3 + \ldots = \frac{1}{1-\alpha}, \ |\alpha| < 1.$$

Using the above formula,

$$X(z) = \frac{1}{1 - az^{-1}}$$

$$= \frac{z}{z-a}.$$

To find the ROC of $X(z)$, we want the condition for which $X(z)$ will converge. This happens when

$$|az^{-1}| < 1$$

$$|a|\frac{1}{|z|} < 1$$

$$|z| > |a|.$$

The ROC is shown in Figure 3.27.

2. $x[n] = -a^n u[-n-1]$. Find the Z transform.

Solution: We will solve this problem by directly applying the analysis equation (3.74).

$$X(z) = \sum_{n=-\infty}^{\infty} x[n]z^{-n}$$

$$= \sum_{n=-\infty}^{\infty} -a^n u[-n-1]z^{-n}$$

133

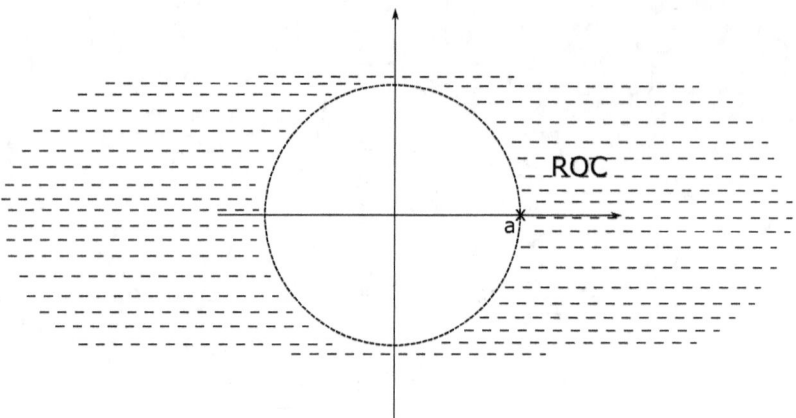

Figure 3.27: ROC: $|z| > |a|$

The signal $y[n] = u[-n-1]$ is shown in Figure 3.28 (refer Section 1.18; obtained from $u[n]$ by first applying time shifting and then time reversal).

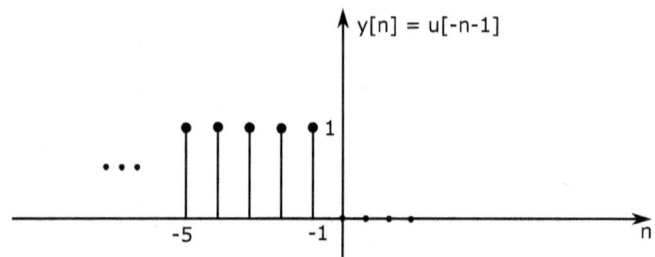

Figure 3.28: Plot of $u[-n - 1]$

Therefore,

$$X(z) = \sum_{n=-\infty}^{-1} -a^n z^{-n}$$

Using a change of variables, let $n = -m$. Thus when $n = -\infty, m = \infty$ and $n = -1, m = 1$. Therefore,

$$X(z) = \sum_{m=\infty}^{1} -a^{-m} z^m$$

$$= -\sum_{m=1}^{\infty} (a^{-1} z)^m$$

This is the sum of infinite terms of a geometric series, which is given by:

$$\alpha + \alpha^2 + \alpha^3 + \ldots = \frac{\alpha}{1-\alpha}, \ |\alpha| < 1.$$

Using the above formula,

$$X(z) = \frac{-a^{-1}z}{1-a^{-1}z}$$
$$= \frac{-z/a}{1-z/a}.$$
$$X(z) = \frac{z}{z-a}$$

To find the ROC of $X(z)$, we want the condition for which $X(z)$ will converge. This happens when

$$|a^{-1}z| < 1$$
$$|a^{-1}||z| < 1$$
$$|z| < |a|.$$

The ROC is shown in Figure 3.29.

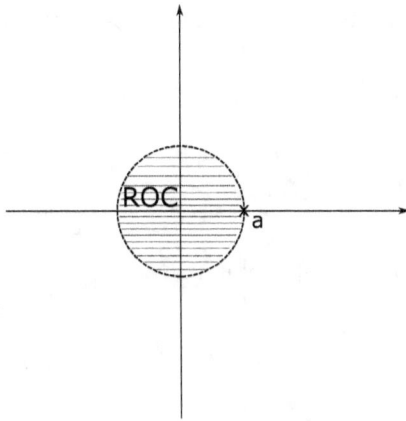

Figure 3.29: ROC: $|z| < |a|$

3.6.4 Properties of the ROC

1. The ROC of the $X(z)$ consists of a ring in the z-plane centered about the origin.

2. The ROC does not contain any poles.

3. If $x[n]$ is of finite duration, then the ROC is the entire z-plane except possibly $z = 0$ and/or $z = \infty$.

4. If $x[n]$ is a right sided sequence and if the circle $|z| = r_0$ is in the ROC, then all the finite values of z for which $|z| > r_0$ will also be in the ROC.

5. If $x[n]$ is a left sided sequence and if the circle $|z| = r_0$ is in the ROC, then all the finite values of z for which $0 < |z| < r_0$ will also be in the ROC.

6. If $x[n]$ is two sided and if the circle $|z| = r_0$ is in the ROC, then the ROC will consist of a ring in the z plane that includes the circle $|z| = r_0$.

7. If the Z transform $X(z)$ of $x[n]$ is rational, then its ROC is bounded by poles or extends to infinity.

8. If the Z transform $X(z)$ of $x[n]$ is rational and if $x[n]$ is right sided then the ROC is the region in the z plane outside the outermost pole. That is, outside the circle of radius equal to the largest magnitude of the poles of $X(z)$.

9. If the Z transform $X(z)$ of $x[n]$ is rational and if $x[n]$ is left sided then the ROC is the region in the z plane inside the innermost pole. That is, inside the circle of radius equal to the smallest magnitude of the poles of $X(z)$.

3.6.5 Solved Examples

1. $x[n] = \delta[n]$. Find the Z transform.

Solution: We will solve this problem by directly applying the analysis equation (3.74).

$$X(z) = \sum_{n=-\infty}^{\infty} x[n]z^{-n}$$

$$= \sum_{n=-\infty}^{\infty} \delta[n]z^{-n}$$

$$= \delta[0]z^0 \quad (\text{only } \delta[n = 0] \text{ is non-zero})$$

$$X(z) = 1.$$

The ROC is the entire z-plane including $z = 0, \infty$ (from property 3 in Section 3.6.4).

> 2. $x[n] = \delta[n-1]$. Find the Z transform.

Solution: We will solve this problem by directly applying the analysis equation (3.74).

$$X(z) = \sum_{n=-\infty}^{\infty} x[n]z^{-n}$$

$$= \sum_{n=-\infty}^{\infty} \delta[n-1]z^{-n}$$

$$= \delta[1-1]z^{-1} \quad (\text{only } \delta[n=1] \text{ is non-zero})$$

$$= \delta[0]z^{-1}$$

$$X(z) = z^{-1} = \frac{1}{z}.$$

The pole of $X(z)$ is located at $z = 0$. Thus, the ROC is the entire z-plane except $z = 0$ and including $z = \infty$ (from property 3 in Section 3.6.4).

> 3. $x[n] = \delta[n+1]$. Find the Z transform.

Solution: We will solve this problem by directly applying the analysis equation (3.74).

$$X(z) = \sum_{n=-\infty}^{\infty} x[n]z^{-n}$$

$$= \sum_{n=-\infty}^{\infty} \delta[n+1]z^{-n}$$

$$= \delta[-1+1]z^{1} \quad (\text{only } \delta[n=-1] \text{ is non-zero})$$

$$= \delta[0]z^{1}$$

$$X(z) = z^{1} = z.$$

The ROC is the entire z-plane except $z = \infty$ (from property 3 in Section 3.6.4).

4. Find the Z transform.

$$x[n] = \begin{cases} a^n, & 0 \leq n \leq N - 1, a > 0 \\ 0, & \text{otherwise} \end{cases}$$

Solution: We will solve this problem by directly applying the analysis equation (3.74).

$$X(z) = \sum_{n=-\infty}^{\infty} x[n]z^{-n}$$

$$= \sum_{n=0}^{N-1} a^n z^{-n} = \sum_{n=0}^{N-1} (az^{-1})^n.$$

This is the sum of N terms of a geometric series, which is given by:

$$1 + \alpha + \alpha^2 + \alpha^3 + \ldots + \alpha^{N-1} = \frac{1 - \alpha^N}{1 - \alpha}.$$

Using the above formula,

$$X(z) = \frac{1 - (az^{-1})^N}{1 - az^{-1}}$$

$$= \frac{1 - a^N z^{-N}}{1 - az^{-1}}$$

$$= \frac{(z^N - a^N)/z^N}{(z - a)/z}$$

$$= \frac{z^N - a^N}{z^N} \frac{z}{z - a}$$

$$= \frac{1}{z^{N-1}} \frac{z^N - a^N}{z - a}$$

From the numerator, there are N zeros for $X(z)$ (since the equation is of the Nth order). The location of the k roots are given by $z_k = ae^{jk(\frac{2\pi}{N})}$, where $k = 0 \ldots (N - 1)$. This is shown in Figure 3.30.

From the denominator, there at N poles for $X(z)$. The location of the poles are given by $z = a$ and $z = 0, (N-1)^{th}$ order. This is also shown in Figure 3.30

We see that there is a pole-zero cancellation at $z = a$. Thus the ROC of $X(z)$ is the entire z-plane except $z = 0$ (from property 3 in Section 3.6.4, note that $x[n]$ is of finite duration).

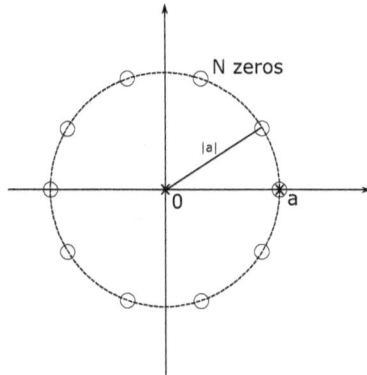

Figure 3.30: Pole zero diagram

3.6.6 Common Z Transform Pairs

Some commonly used Z Transform pairs are given in Table 3.5. For a more elaborate list of common Z Transform pairs, please refer [1].

3.6.7 Properties of the Z transform

Consider a signal $x[n]$ with Z Transform $X(z)$, that is,

$$x(t) \overset{Z}{\leftrightarrow} X(z), \text{ ROC: } R$$

We are interested to know what happens to the Z Transform when various signal operations are done in the time (n) domain.

Some of the major properties of the Z Transform that are often used when solving numberical problems are discussed next; for a full list of the properties please refer [1].

Linearity

Consider two signals $x[n]$ and $y[n]$ with

$$x[n] \overset{Z}{\leftrightarrow} X(z), \text{ ROC: } R_1$$
$$y[n] \overset{Z}{\leftrightarrow} Y(z), \text{ ROC: } R_2$$

Then

$$z[n] = Ax[n] + By[n] \overset{LT}{\longleftrightarrow} Z(z) = AX(z) + BY(z), \qquad (3.75)$$

where $Z(z)$ is the Laplace Transform of $z[n]$. The ROC of $Z(z)$ is: at least $R_1 \cap R_2$. Therefore, *linearity in the time domain* corresponds to *linearity in the z domain*.

Signal	Z Transform	ROC
$\delta[n]$	1	All z
$u[n]$	$\frac{1}{1-z^{-1}}$	$|z| > 1$
$-u[-n-1]$	$\frac{1}{1-z^{-1}}$	$|z| < 1$
$\alpha^n u[n]$	$\frac{1}{1-\alpha z^{-1}}$	$|z| > |\alpha|$
$-\alpha^n u[-n-1]$	$\frac{1}{1-\alpha z^{-1}}$	$|z| < |\alpha|$
$n\alpha^n u[n]$	$\frac{\alpha z^{-1}}{(1-\alpha z^{-1})^2}$	$|z| > |\alpha|$
$-n\alpha^n u[-n-1]$	$\frac{\alpha z^{-1}}{(1-\alpha z^{-1})^2}$	$|z| < |\alpha|$

Table 3.5: Common Z Transform pairs

Time Shifting

Consider $x[n]$ with

$$x[n] \overset{Z}{\leftrightarrow} X(z), \text{ ROC: } R$$

Then

$$x[n - n_0] \overset{Z}{\leftrightarrow} z^{-n_0} X(z) \tag{3.76}$$

The ROC is: R, except for the possible addition or deletion of the origin.

Time Reversal

Consider $x[n]$ with

$$x[n] \overset{Z}{\leftrightarrow} X(z), \text{ ROC: } R$$

Then

$$x[-n] \overset{Z}{\leftrightarrow} X(z^{-1}), \text{ ROC: } R^{-1}(\text{the set of points } z^{-1}, \text{ where } z \text{ is in } R) \tag{3.77}$$

Convolution

Consider two signals $x[n]$ and $y[n]$ with

$$x[n] \overset{Z}{\leftrightarrow} X(z), \text{ ROC: } R_1$$
$$y[n] \overset{Z}{\leftrightarrow} Y(z), \text{ ROC: } R_2$$

Then

$$z[n] = x[n] * y[n] \overset{Z}{\leftrightarrow} Z(z) = X(z)Y(z), \tag{3.78}$$

where $Z(z)$ is the Z Transform of $z[n]$. The ROC of $Z(z)$ is: at least $R_1 \cap R_2$.

Therefore, *convolution in the time domain* corresponds to *multiplication in the z domain*.

Conjugation

Consider $x[n]$ with

$$x[n] \overset{Z}{\leftrightarrow} X(z), \text{ ROC: } R$$

Then

$$x^*[n] \overset{Z}{\leftrightarrow} X^*(z^*), \text{ ROC: } R \tag{3.79}$$

First Difference

Consider $x[n]$ with

$$x[n] \overset{Z}{\leftrightarrow} X(z), \text{ ROC: } R$$

Then

$$x[n] - x[n-1] \overset{Z}{\leftrightarrow} (1 - z^{-1})X(z), \text{ ROC: at least the intersection of } R \text{ and } |z| > 0 \tag{3.80}$$

Differentiation in z Domain

Consider $x[n]$ with

$$x[n] \overset{Z}{\leftrightarrow} X(z), \text{ ROC: } R$$

Then

$$nx[n] \overset{Z}{\leftrightarrow} -z\frac{d}{dz}X(z), \text{ ROC: } R \tag{3.81}$$

3.6.8 Solved Examples

1. Given $x[n] = u[-n]$. Find $X(z)$.

Solution: From the known Z transforms (Section 3.6.6),

$$a^n u[n] \overset{Z}{\longleftrightarrow} \frac{1}{1 - az^{-1}}, \ |z| > |a|$$

$$y[n] = u[n] \overset{Z,a=1}{\longleftrightarrow} Y(z) = \frac{1}{1 - z^{-1}}, \ |z| > 1$$

Using the time reversal property,

$$u[-n] = y[-n] \overset{Z}{\longleftrightarrow} Y(1/z), \ |1/z| > 1$$

$$\overset{Z}{\longleftrightarrow} \frac{1}{1 - (\frac{1}{z})^{-1}}, \ |z| < 1$$

$$u[-n] \overset{Z}{\longleftrightarrow} \frac{1}{1 - z}, \ |z| < 1.$$

2. $x[n] = 7(1/3)^n u[n] - 6(1/2)^n u[n]$. Find the Z transform.

Solution: From the known Z transforms (Section 3.6.6),

$$a^n u[n] \overset{Z}{\longleftrightarrow} \frac{1}{1 - az^{-1}}, \ |z| > |a|$$

$$\left(\frac{1}{3}\right)^n u[n] \overset{Z,a=1/3}{\longleftrightarrow} \frac{1}{1 - \frac{1}{3}z^{-1}}, \ |z| > 1/3$$

$$\left(\frac{1}{2}\right)^n u[n] \overset{Z,a=1/2}{\longleftrightarrow} \frac{1}{1 - \frac{1}{2}z^{-1}}, \ |z| > 1/2$$

$$(3.82)$$

Thus, from the linearity property of the transform,

$$X(z) = 7\frac{1}{1 - \frac{1}{3}z^{-1}} - 6\frac{1}{1 - \frac{1}{2}z^{-1}}, \ |z| > 1/2$$

$$= 7\frac{z}{z - 1/3} - 6\frac{z}{z - 1/2}$$

$$X(z) = \frac{z(z - 3/2)}{(z - 1/3)(z - 1/2)}, \ |z| > 1/2.$$

142

The ROC is shown in Figure 3.31 [3].

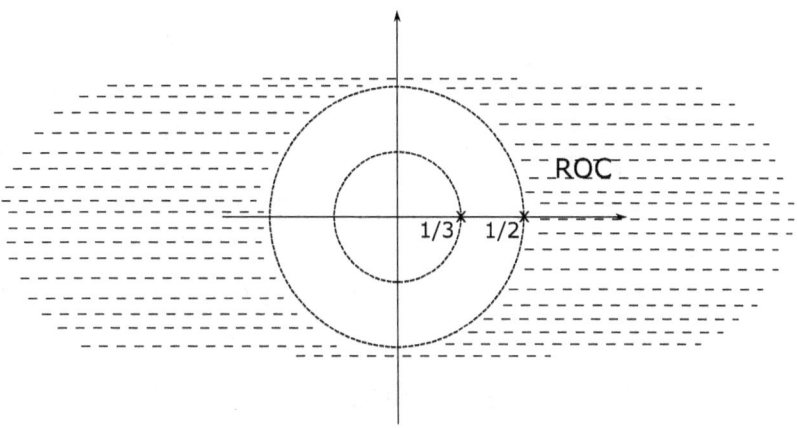

Figure 3.31: ROC: $|z| > 1/2$

---

3. Find x[n], given:

$$X(z) = \frac{3 - 5/6z^{-1}}{(1 - 1/4z^{-1})(1 - 1/3z^{-1})}, |z| > 1/3$$

Solution:

$$X(z) = \frac{3 - 5/6z^{-1}}{(1 - 1/4z^{-1})(1 - 1/3z^{-1})}$$

$$= \frac{(3z - 5/6)z}{(z - 1/4)(z - 1/3)}, \ |z| > 1/3$$

Thus the poles are at $z = 1/4, 1/3$. The ROC is shown in Figure 3.32.

$$X(z) = \frac{A}{(1 - 1/4z^{-1})} + \frac{B}{(1 - 1/3z^{-1})}$$

Solving for A, B gives $A = 1, B = 2$. Thus,

$$X(z) = \frac{1}{(1 - 1/4z^{-1})} + \frac{2}{(1 - 1/3z^{-1})}, \ |z| > 1/3$$

---

[3]The common region for $|z| > 1/3$ and $|z| > 1/2$ is $|z| > 1/2$.

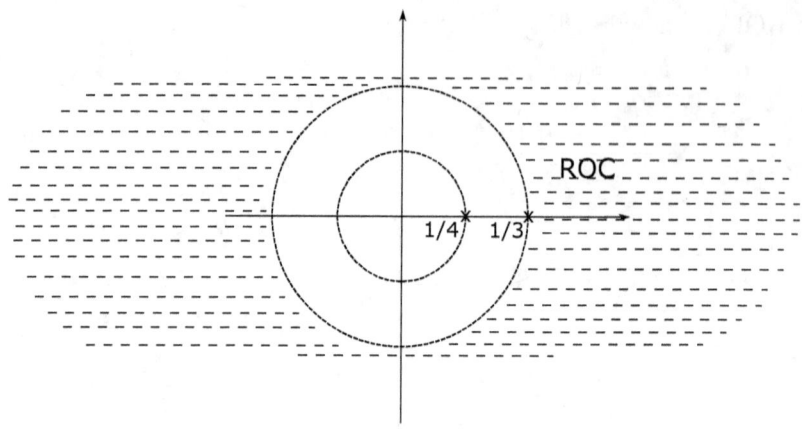

Figure 3.32: ROC: $|z| > 1/3$

From the known Z transforms (Section 3.6.6),

$$\frac{1}{(1 - 1/4z^{-1})}, \ |z| > 1/4 \xleftrightarrow{Z, a=1/4} (\frac{1}{4})^n u[n]$$

$$\frac{1}{(1 - 1/3z^{-1})}, \ |z| > 1/3 \xleftrightarrow{Z, a=1/3} (\frac{1}{3})^n u[n]$$

Therefore,

$$x[n] = (\frac{1}{4})^n u[n] + (\frac{1}{3})^n u[n].$$

---

4. Find x[n], given:

$$X(z) = \frac{3 - 5/6z^{-1}}{(1 - 1/4z^{-1})(1 - 1/3z^{-1})}, 1/4 < |z| < 1/3$$

Solution: From the previous problem,

$$X(z) = \frac{1}{(1 - 1/4z^{-1})} + \frac{2}{(1 - 1/3z^{-1})}, \ 1/4 < |z| < 1/3$$

The ROC is shown in Figure 3.33.

From the known Z transforms (Section 3.6.6),

$$\frac{1}{(1 - 1/4z^{-1})}, \ |z| > 1/4 \xleftrightarrow{Z, a=1/4} (\frac{1}{4})^n u[n]$$

$$\frac{1}{(1 - 1/3z^{-1})}, \ |z| < 1/3 \xleftrightarrow{Z, a=1/3} -(\frac{1}{3})^n u[-n - 1]$$

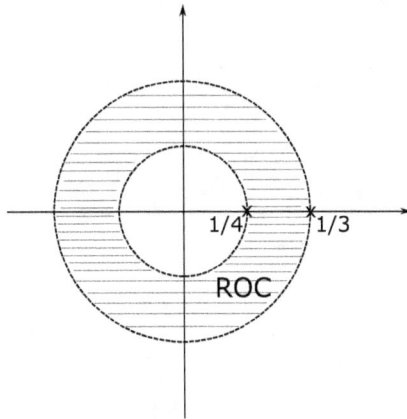

Figure 3.33: ROC: $1/4 < |z| < 1/3$

Therefore,

$$x[n] = (\frac{1}{4})^n u[n] - 2(\frac{1}{3})^n u[-n-1].$$

5. Find x[n], given:

$$X(z) = \frac{3 - 5/6z^{-1}}{(1 - 1/4z^{-1})(1 - 1/3z^{-1})}, |z| < 1/4$$

Solution: From the previous problem,

$$X(z) = \frac{1}{(1 - 1/4z^{-1})} + \frac{2}{(1 - 1/3z^{-1})}, \ |z| < 1/4$$

The ROC is shown in Figure 3.34.

From the known Z transforms (Section 3.6.6),

$$\frac{1}{(1 - 1/4z^{-1})}, \ |z| < 1/4 \xleftrightarrow{Z, a=1/4} -(\frac{1}{4})^n u[-n-1]$$

$$\frac{1}{(1 - 1/3z^{-1})}, \ |z| < 1/3 \xleftrightarrow{Z, a=1/3} -(\frac{1}{3})^n u[-n-1]$$

Therefore,

$$x[n] = -(\frac{1}{4})^n u[-n-1] - 2(\frac{1}{3})^n u[-n-1].$$

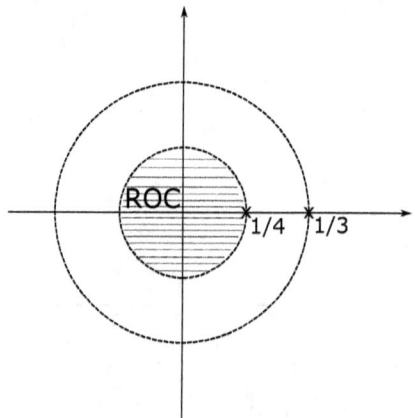

Figure 3.34: ROC: $|z| < 1/4$

4 Analysing LTI Systems Using Transforms

An LTI system refers to a system that satisfies the linearity and time-invariance properties. This was discussed in Section 2.8. From the properties of the CTFT (Section 3.2.2), Laplace Transform (Section 3.3.6), DTFT (Section 3.5.3) and the Z Transform (Section: 3.6.7), we understand that convolution in the time-domain leads to multiplication in the frequency (or s or z) domain. In this chapter, it is this property of the transform that is utilized in analyzing an LTI system. The concept map for the topics covered in this chapter is shown in Figure 4.1.

4.1 Analysing CT LTI Systems Using CTFT

Recall that the output from a continuous-time (CT) LTI system with input $x(t)$ and impulse response $h(t)$ is given by $y(t) = x(t) * h(t)$. Using the CTFT properties the Fourier Transform of $y(t)$ is $Y(j\omega) = X(j\omega)H(j\omega)$. $H(j\omega)$ is called the *frequency response* of the LTI system. Also recall that a CT LTI system can be described using a linear constant coefficient differential equation with the input $x(t)$ on the RHS of the equation and the output $y(t)$ on the LHS of the equation.

The goal of this section is to derive an expression for the frequency response $H(j\omega)$ and hence $h(t)$ of the LTI system given the linear constant coefficient differential equation.

4.1.1 Deriving the Frequency Response

A CT LTI system described by a linear constant coefficient differential equation is given by:

$$a_0 y(t) + a_1 \frac{dy(t)}{dt} + a_2 \frac{d^2 y(t)}{dt^2} + \cdots + a_N \frac{d^N y(t)}{dt^N} =$$
$$b_0 x(t) + b_1 \frac{dx(t)}{dt} + b_2 \frac{d^2 x(t)}{dt^2} + \cdots + b_M \frac{d^M x(t)}{dt^M} \tag{4.1}$$

Note the following:

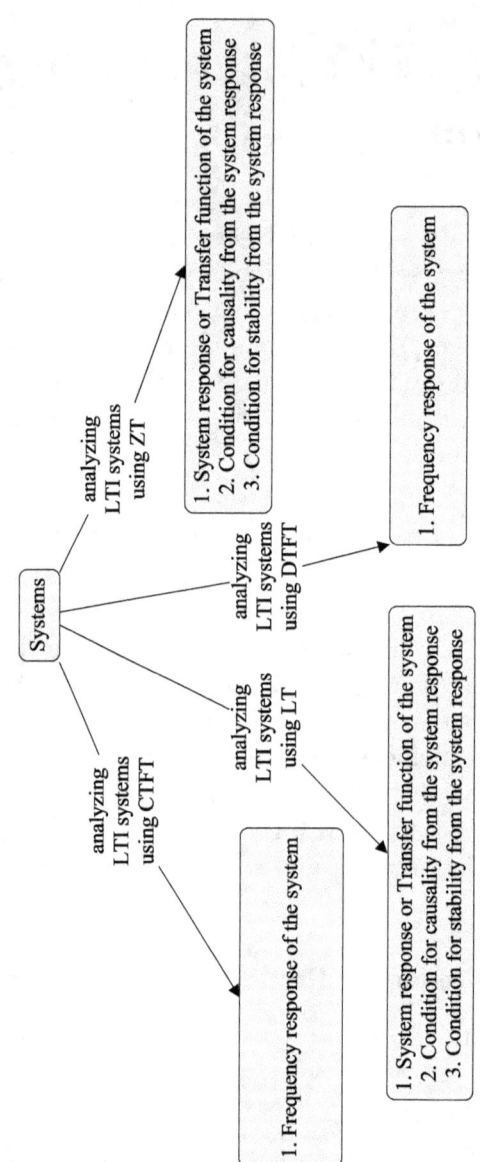

Figure 4.1: Concept map of topics in Chapter 4

1. The input of the system $(x(t))$ is on the RHS of the equation.

2. The output of the system $(y(t))$ is on the LHS of the equation.

3. The coefficients of the input $x(t)$ are $b_0, b_1, b_2, \cdots, b_M$.

4. The coefficients of the output $y(t)$ are $a_0, a_1, a_2, \cdots, a_N$.

5. The order of the RHS of the equation is M.

6. The order of the LHS of the equation is N.

The above equation can be written as follows:

$$\sum_{k=0}^{k=N} a_k \frac{d^k y(t)}{dt^k} = \sum_{k=0}^{k=M} b_k \frac{d^k x(t)}{dt^k}. \tag{4.2}$$

Taking the Fourier Transform on both sides of the above equation,

$$FT\left\{\sum_{k=0}^{k=N} a_k \frac{d^k y(t)}{dt^k}\right\} = FT\left\{\sum_{k=0}^{k=M} b_k \frac{d^k x(t)}{dt^k}\right\}. \tag{4.3}$$

Utilizing the linearity property of the Fourier Transform,

$$\sum_{k=0}^{k=N} a_k \, FT\left\{\frac{d^k y(t)}{dt^k}\right\} = \sum_{k=0}^{k=M} b_k \, FT\left\{\frac{d^k x(t)}{dt^k}\right\}. \tag{4.4}$$

From the differentiation property of the Fourier Transform,

$$\frac{d}{dt} \longleftrightarrow (j\omega) \tag{4.5}$$

and

$$\frac{d^k}{dt^k} \longleftrightarrow (j\omega)^k. \tag{4.6}$$

Applying this, Equation (4.4) becomes:

$$\sum_{k=0}^{k=N} a_k (j\omega)^k Y(j\omega) = \sum_{k=0}^{k=M} b_k (j\omega)^k X(j\omega) \tag{4.7}$$

$$Y(j\omega)\left\{\sum_{k=0}^{k=N} a_k (j\omega)^k\right\} = X(j\omega)\left\{\sum_{k=0}^{k=M} b_k (j\omega)^k\right\}. \tag{4.8}$$

Therefore,

$$H(j\omega) = \frac{Y(j\omega)}{X(j\omega)} = \frac{\sum_{k=0}^{k=M} b_k (j\omega)^k}{\sum_{k=0}^{k=N} a_k (j\omega)^k}. \tag{4.9}$$

Note that $H(j\omega)$ is a ratio of two polynomials in $j\omega$. Once $H(j\omega)$ is known, the impulse response $h(t)$ can be calculated using the inverse Fourier Transform.

4.1.2 Solved Examples

---

1. Given a system with the following input-output relation

$$\frac{d}{dt}y(t) + ay(t) = x(t), a > 0.$$

Find the impulse response of the system.

Solution: Comparing the given differential equation with Equation (4.1),

$$N = 1 \qquad\qquad M = 0$$
$$a_0 = a \qquad\qquad b_0 = 1$$
$$a_1 = 1$$

From Equation (4.9),

$$H(j\omega) = \frac{1}{a + j\omega}, a > 0.$$

From the known transforms in Section 3.2.1,

$$h(t) = e^{-at}u(t).$$

---

2. Given a system with the following input-output relation

$$\frac{d^2}{dt^2}y(t) + 4\frac{d}{dt}y(t) + 3y(t) = \frac{d}{dt}x(t) + 2x(t)$$

Find the impulse response of the system.

Solution: Comparing the given differential equation with Equation (4.1),

$$N = 1 \qquad\qquad M = 1$$
$$a_0 = a \qquad\qquad b_0 = 1$$
$$a_1 = 1$$

From Equation (4.9),

$$H(j\omega) = \frac{2 + j\omega}{3 + 4(j\omega) + (j\omega)^2}$$

$$= \frac{2 + j\omega}{(j\omega + 1)(j\omega + 3)}$$

$$= \frac{A}{1 + j\omega} + \frac{B}{3 + j\omega}$$

Solving for A, B gives $A = 1/2, B = 1/2$.

Therefore,

$$H(j\omega) = \frac{1/2}{1 + j\omega} + \frac{1/2}{3 + j\omega}$$

From the known transforms in Section 3.2.1,

$$\frac{1/2}{1 + j\omega} \xleftrightarrow{\text{CTFT},a=1} \frac{1}{2}e^{-t}u(t)$$

$$\frac{1/2}{3 + j\omega} \xleftrightarrow{\text{CTFT},a=3} \frac{1}{2}e^{-3t}u(t)$$

Therefore

$$h(t) = \frac{1}{2}e^{-t}u(t) + \frac{1}{2}e^{-3t}u(t).$$

4.2 Analysing CT LTI Systems Using Laplace Transform

Recall that the output from a continuous-time LTI system with input $x(t)$ and impulse response $h(t)$ is $y(t) = x(t) * h(t)$. Using the Laplace Transform properties, the Laplace Transform of $y(t)$ is $Y(s) = X(s)H(s)$. $H(s)$ is called the *system response* or *transfer function* of the LTI system. Also recall that an LTI system can be described using a linear constant coefficient differential equation with the input $x(t)$ on the RHS of the equation and the output $y(t)$ on the LHS of the equation.

The goal of this section is to derive an expression for the system response $H(s)$ and hence $h(t)$ of the LTI system given the linear constant coefficient differential equation.

4.2.1 Deriving the System Response or Transfer Function $H(s)$

In Section 4.1.1, the frequency response ($H(j\omega)$) of a LTI system from a linear constant coefficient differential equation was derived. Using a similar approach, the system response ($H(s)$) of a LTI system can also be derived from the linear constant coefficient differential equation.

A CT LTI system described by a linear constant coefficient differential equation is given by:

$$a_0 y(t) + a_1 \frac{dy(t)}{dt} + a_2 \frac{d^2 y(t)}{dt^2} + \cdots + a_N \frac{d^N y(t)}{dt^N} =$$
$$b_0 x(t) + b_1 \frac{dx(t)}{dt} + b_2 \frac{d^2 x(t)}{dt^2} + \cdots + b_M \frac{d^M x(t)}{dt^M} \tag{4.10}$$

The expression for $H(s)$ is given by:

$$H(s) = \frac{Y(s)}{X(s)} = \frac{\sum_{k=0}^{k=M} b_k(s)^k}{\sum_{k=0}^{k=N} a_k(s)^k}. \tag{4.11}$$

Note the following:

1. The coefficients of the input $x(t)$ are $b_0, b_1, b_2, \cdots, b_M$.

2. The coefficients of the output $y(t)$ are $a_0, a_1, a_2, \cdots, a_N$.

Also note that $H(s)$ is a ratio of two polynomials in s and that this method tells us nothing about the ROC of $H(s)$. Once $H(s)$ is known and the ROC is also known (or given in the problem), the impulse response $h(t)$ can be calculated using the inverse Laplace Transform.

4.2.2 Solved Example

1. Given:
$$\frac{dy(t)}{dt} + 3y(t) = x(t).$$
Find the transfer function.

Solution: Comparing the given differential equation with Equation (4.10),

$$N = 1 \qquad\qquad M = 0$$
$$a_0 = 3 \qquad\qquad b_0 = 1$$
$$a_1 = 1$$

From Equation (4.11),

$$H(s) = \frac{1}{3 + s} = \frac{1}{s + 3}.$$

4.2.3 Condition for Causality from the Transfer Function $H(s)$

In this section, we are trying to answer the following question: what conditions should $H(s)$ satisfy, if a LTI system needs to be causal?

To answer this, recall from Section 2.17, that the condition for a LTI system to be causal (in the time domain) is given by: $h(t) = 0$ for $t < 0$. That is, the signal $h(t)$ is right-sided. This implies that:

1. The ROC associated with $H(s)$ for a causal system is the right half plane (from Property 4 of the ROC, refer Section 3.3.4).

2. Additionaly, from Property 8 of the ROC, for a system with *rational* $H(s)$, causality of the system implies the ROC being the right half plane to the right of the rightmost pole.

4.2.4 Condition for Stability from the Transfer Function $H(s)$

In this section, we are trying to answer the following question: what conditions should $H(s)$ satisfy, if a LTI system needs to be stable?

To answer this, recall from Section 2.18, that the condition for a LTI system to be stable (in the time domain) is given by: $\int_{-\infty}^{\infty} |h(t)|dt < 0$. That is, $h(t)$ is absolutely integrable and hence the Fourier Transform of $h(t)$ converges. Therefore the ROC of $H(s)$ must include the entire $j\omega$ axis. This implies that:

1. An LTI system is stable if and only if the ROC of $H(s)$ includes the entire $j\omega$ axis.

2. Additionaly, if $H(s)$ is rational and the system is causal, then the system is stable if and only if all the poles of $H(s)$ lie in the left half plane. That is, all the poles have a negative real part.

4.2.5 Solved Examples

1. An LTI system has $h(t) = e^{-t}u(t)$. Is the system stable, causal?

Solution: From the list of known transforms (Section 3.3.5),

$$h(t) = e^{-t}u(t) \xrightarrow{\text{LT},\alpha=1} H(s) = \frac{1}{s+1}, \quad Re(s) > -1.$$

The ROC is indicated in Figure 4.2.

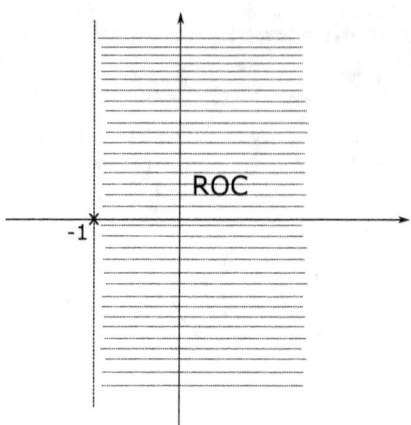

Figure 4.2: ROC: $Re(s) > -1$

Causality check: $H(s)$ is a rational function and since the ROC is the right of the rightmost pole, from Section 4.2.3, the system is said to be causal.

Stability check: Since the ROC includes the $j\omega$ axis, from Section 4.2.4, the system is said to be stable.

---

2. An LTI system has $h(t) = e^{2t}u(t)$. Is the system stable?

Solution: From the list of known transforms (Section 3.3.5),

$$h(t) = e^{2t}u(t) \xleftrightarrow{\text{LT},\alpha=-2} H(s) = \frac{1}{s-2}, \quad Re(s) > 2.$$

The ROC is indicated in Figure 4.3.

Stability check: Since the ROC does not include the $j\omega$ axis, from Section 4.2.4, the system is said to be not stable.

---

3. Given: $H(s) = \frac{s-1}{(s+1)(s-2)}$. Indicate the ROC if: (i) causal, (ii) stable, (iii) if not causal and unstable.

Solution:

The poles of $H(s)$ are located at $s = -1, s = 2$.

(i) If causal: From Section 4.2.3, for a rational $H(s)$, the ROC must be to the right of the rightmost pole. Therefore the ROC is $Re(s) > 2$. This is shown in Figure 4.4.

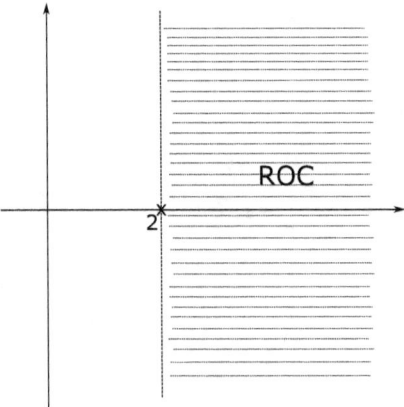

Figure 4.3: ROC: $Re(s) > 2$

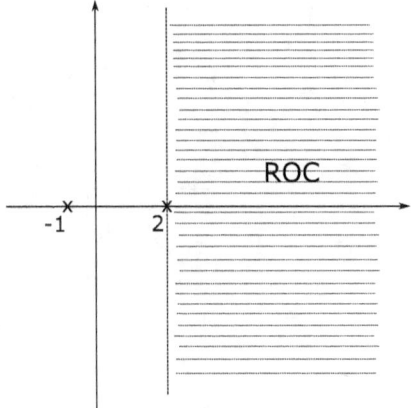

Figure 4.4: ROC: $Re(s) > 2$

(ii) <u>If stable</u>: From Section 4.2.4, the ROC must include the $j\omega$ axis. Therefore the ROC is $-1 < Re(s) < 2$. [1] This is shown in Figure 4.5.

(iii) <u>If not causal and unstable</u>: In this case, the ROC will be to the left half plane (since the system is not causal) and the ROC should not include the $j\omega$ axis. Therefore the ROC that satisfies both these conditions is $Re(s) < -1$. This is shown in Figure 4.6.

---

[1] Note that the other possible ROCs are $Re(s) > -1$ and $Re(s) < 2$. Both of these are not possible, since the ROC will include a pole, which violates property 2 in Section 3.3.4.

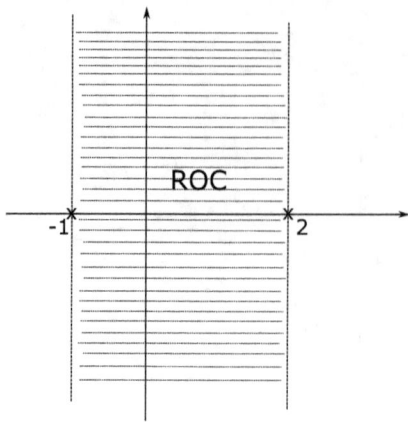

Figure 4.5: ROC: $-1 < Re(s) < 2$

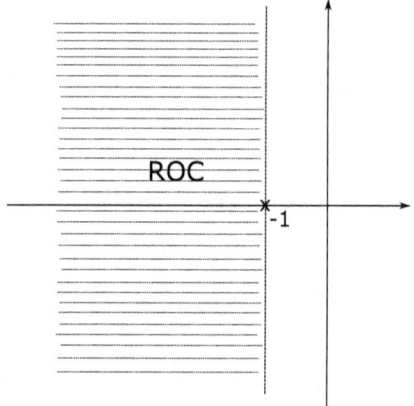

Figure 4.6: ROC: $Re(s) < -1$

4. Given $x(t) = e^{-3t}u(t), y(t) = (e^{-t} - e^{-2t})u(t)$. Is the system causal, stable?

Solution: To determine if the system is causal or stable, first $H(s)$ and its ROC needs to be determined. $H(s)$ is given by $\frac{Y(s)}{X(s)}$.

From the list of known transforms (Section 3.3.5),

$$x(t) = e^{-3t}u(t) \xleftrightarrow{\text{LT},\alpha=3} X(s) = \frac{1}{s+3}, \quad Re(s) > -3.$$

The ROC is indicated in Figure 4.7.

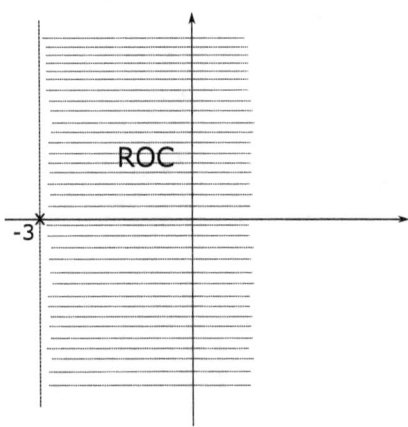

Figure 4.7: ROC of $X(s)$: $Re(s) > -3$

$$y(t) = e^{-t}u(t) - e^{-2t}u(t)$$

$$e^{-t}u(t) \xleftrightarrow{\text{LT},\alpha=1} \frac{1}{s+1}, \quad Re(s) > -1$$
$$e^{-2t}u(t) \xleftrightarrow{\text{LT},\alpha=2} \frac{1}{s+2}, \quad Re(s) > -2$$

Therefore,

$$Y(s) = \frac{1}{s+1} - \frac{1}{s+2}, \quad Re(s) > -1$$
$$Y(s) = \frac{1}{(s+1)(s+2)}, \quad Re(s) > -1.$$

The ROC is indicated in Figure 4.9.

Therefore

$$H(s) = \frac{Y(s)}{X(s)}$$
$$= \frac{s+3}{(s+1)(s+2)}.$$

Since the ROC of $H(s)$ is not given, there are three options for the ROC of $H(s)$: (i) $Re(s) < -2$, (ii) $-2 < Re(s) < -1$, (ii) $Re(s) > -1$. To determine the ROC of $H(s)$, recall from the convolution property of the Laplace

157

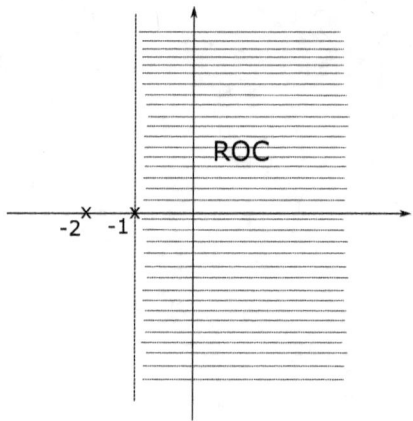

Figure 4.8: ROC of $Y(s)$: $Re(s) > -1$

Transform (Section 3.3.6),

$$Y(s) = X(s)H(s)$$
$$\text{ROC of } Y(s) = \text{ROC of } X(s) \cap \text{ROC of } H(s).$$

Since the ROC of $X(s)$ and $Y(s)$ are known, the only option for the ROC of $H(s)$ that satisfies the above relation is $Re(s) > -1$. This is shown in Figure 4.9.

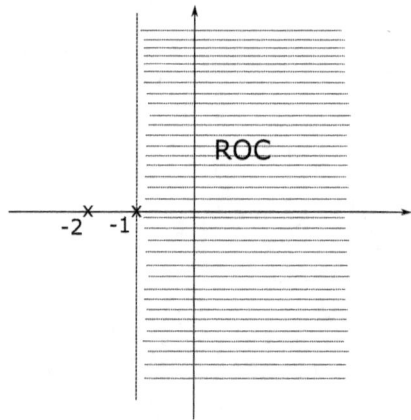

Figure 4.9: ROC of $H(s)$: $Re(s) > -1$

Since the ROC is to the right of the rightmost pole and includes the $j\omega$ axis, the system is causal and stable.

4.3 Analysing DT LTI Systems Using DTFT

Recall that the output from a discrete-time LTI system with input $x[n]$ and impulse response $h[n]$ is $y[n] = x[n]*h[n]$. Using the DTFT properties, the Fourier Transform of $y[n]$ is $Y(e^{j\omega}) = X(e^{j\omega})H(e^{j\omega})$. $H(e^{j\omega}) = Y(e^{j\omega})/X(e^{j\omega})$ is called the *frequency response* of the DT LTI system. Also recall that a DT LTI system can be described using a linear constant coefficient difference equation with the input $x[n]$ on the RHS of the equation and the output $y[n]$ on the LHS of the equation.

The goal of this section is to derive an expression for the frequency response $H(e^{j\omega})$ and hence $h[n]$ of the DT LTI system given the linear constant coefficient difference equation.

4.3.1 Deriving the Frequency Response

A DT LTI system described by a linear constant coefficient difference equation is given by:

$$a_0 y[n] + a_1 y[n-1] + a_2 y[n-2] + \cdots + a_N y[n-N] =$$
$$b_0 x[n] + b_1 x[n-1] + b_2 x[n-2] + \cdots + b_M x[n-M] \tag{4.12}$$

Note the following:

1. The input of the system ($x[n]$) is on the RHS of the equation.

2. The output of the system ($y[n]$) is on the LHS of the equation.

3. The coefficients of the input $x[n]$ are $b_0, b_1, b_2, \cdots, b_M$.

4. The coefficients of the output $y[n]$ are $a_0, a_1, a_2, \cdots, a_N$.

5. The order of the RHS of the equation is M.

6. The order of the LHS of the equation is N.

The above equation can be written as follows:

$$\sum_{k=0}^{k=N} a_k y[n-k] = \sum_{k=0}^{k=M} b_k x[n-k] \tag{4.13}$$

Taking the DTFT on both sides of the above equation, then applying the linearity and differencing in time properties of the DTFT, it can be shown that:

$$H(e^{j\omega}) = \frac{Y(e^{j\omega})}{X(e^{j\omega})} = \frac{\sum_{k=0}^{k=M} b_k (e^{-jk\omega})}{\sum_{k=0}^{k=N} a_k (e^{-jk\omega})}. \tag{4.14}$$

Note that $H(e^{j\omega})$ is a ratio of two polynomials in $e^{j\omega}$. Once $H(e^{j\omega})$ is known, the impulse response $h[n]$ can be calculated using the inverse Fourier Transform.

4.3.2 Solved Examples

1. Consider a LTI system described by $y[n] - ay[n-1] = x[n]$. Find $h[n]$.

Solution: Comparing the given difference with Equation (4.12),

$$N = 1 \qquad\qquad M = 0$$
$$a_0 = 1 \qquad\qquad b_0 = 1$$
$$a_1 = -a$$

From Equation (4.14),

$$H(e^{j\omega}) = \frac{Y(e^{j\omega})}{X(e^{j\omega})}$$
$$= \frac{1}{1 + (-a)e^{-j\omega}}$$
$$= \frac{1}{1 - ae^{-j\omega}} \qquad\qquad (4.15)$$

From the known DT Fourier Transforms (Section (3.5.2)),

$$h[n] = a^n u[n].$$

2. Consider a LTI system described by $y[n] - 3/4y[n-1] + 1/8y[n-2] = 2x[n]$. Find $h[n]$.

Solution: Comparing the given difference with Equation (4.12),

$$N = 2 \qquad\qquad M = 0$$
$$a_0 = 1 \qquad\qquad b_0 = 2$$
$$a_1 = -3/4$$
$$a_2 = 1/8$$

From Equation (4.14),

$$H(e^{j\omega}) = \frac{Y(e^{j\omega})}{X(e^{j\omega})}$$

$$= \frac{2}{1 + (-3/4)e^{-j\omega} + (1/8)e^{-j2\omega}}$$

$$= \frac{2}{(1 - \frac{e^{-j\omega}}{4})(1 - \frac{e^{-j\omega}}{2})}$$

$$= \frac{A}{(1 - \frac{e^{-j\omega}}{4})} + \frac{B}{(1 - \frac{e^{-j\omega}}{2})} \qquad (4.16)$$

Solving for A and B, $A = -2$ and $B = 4$.

Therefore

$$H(z) = \frac{4}{(1 - \frac{e^{-j\omega}}{2})} - \frac{2}{(1 - \frac{e^{-j\omega}}{4})}.$$

Using the known DT Fourier Transforms (Section (3.5.2)),

$$h[n] = 4(\frac{1}{2})^n u[n] - 2(\frac{1}{4})^n u[n].$$

4.4 Analysing DT LTI Systems Using Z Transform

Recall that the output from a discrete-time (DT) LTI system with input $x[n]$ and impulse response $h[n]$ is $y[n] = x[n] * h[n]$. Using the Z Transform properties, the Z Transform of $y[n]$ is $Y(z) = X(z)H(z)$. $H(z)$ is called the *system response* or *transfer function* of the DT LTI system. Also recall that an LTI system can be described using a linear constant coefficient difference equation with the input $x[n]$ on the RHS of the equation and the output $y[n]$ on the LHS of the equation.

The goal of this section is to derive an expression for the system response $H(z)$ and hence $h[n]$ of the LTI system given the linear constant coefficient difference equation.

4.4.1 Deriving the System Response or Transfer Function $H(z)$

In Section 4.3.1, the frequency response ($H(e^{j\omega})$) of a DT LTI system from a linear constant coefficient difference equation was derived. Using a similar approach, the system response ($H(z)$) of a DT LTI system can also be derived from the linear constant coefficient difference equation.

A DT LTI system described by a linear constant coefficient difference equation is given by:

$$a_0 y[n] + a_1 y[n-1] + a_2 y[n-2] + \cdots + a_N y[n-N] =$$
$$b_0 x[n] + b_1 x[n-1] + b_2 x[n-2] + \cdots + b_M x[n-M] \tag{4.17}$$

The above equation can be written as follows:

$$\sum_{k=0}^{k=N} a_k y[n-k] = \sum_{k=0}^{k=M} b_k x[n-k] \tag{4.18}$$

The expression for $H(z)$ can be derived and is given by:

$$H(z) = \frac{Y(z)}{X(z)} = \frac{\sum_{k=0}^{k=M} b_k (z^{-k})}{\sum_{k=0}^{k=N} a_k (z^{-k})} \tag{4.19}$$

Note the following:

1. The coefficients of the input $x[n]$ are $b_0, b_1, b_2, \cdots, b_M$.

2. The coefficients of the output $y[n]$ are $a_0, a_1, a_2, \cdots, a_N$.

Also note that $H(z)$ is a ratio of two polynomials in z and that this method tells us nothing about the ROC of $H(z)$. Once $H(z)$ is known and the ROC is also known (or given in the problem), the impulse response $h[n]$ can be calculated using the inverse Z Transform.

4.4.2 Condition for Causality from the Transfer Function $H(z)$

In this section, we are trying to answer the following question: what conditions should $H(z)$ satisfy, if a LTI system needs to be causal?

To answer this, recall from Section 2.17, that the condition for a LTI system to be causal (in the time domain) is given by: $h[n] = 0$ for $n < 0$. That is, the signal $h[n]$ is right-sided. This implies that:

1. A DT LTI system is causal iff the ROC of $H(z)$ is the exterior of a circle including infinity (from Property 4 of the ROC, refer Section 3.6.4).

2. Additionaly, from Property 8 of the ROC, a system with a *rational* $H(z)$ is causal iff:

    a) ROC is the exterior of a circle outside the outermost pole, and

    b) with $H(z)$ expressed a ratio of polynomials in z, the order of the numerator cannot be greater than the order of the denominator.

4.4.3 Condition for Stability from the Transfer Function $H(z)$

In this section, we are trying to answer the following question: what conditions should $H(z)$ satisfy, if a LTI system needs to be stable?

To answer this, recall from Section 2.18, that the condition for a DT LTI system to be stable (in the time domain) is given by: $\sum_{n=-\infty}^{\infty} |h[n]| < \infty$. That is, $h[n]$ is absolutely integrable and hence the Fourier Transform of $h[n]$ converges. Therefore the ROC of $H(z)$ must include the unit circle. This implies that:

1. An LTI system is stable if and only if the ROC of $H(s)$ includes the unit circle $|z| = 1$.

2. Additionally, if $H(z)$ is rational and the system is causal, then the system is stable iff all the poles of $H(z)$ lie inside the unit circle, that is, they must all have magnitude smaller than 1.

4.4.4 Solved Examples

1. Given:
$$H(z) = \frac{z^3 - 2z + z}{z^2 + 1/4z + 1/8}.$$

Is the system causal?

Solution: From Section 4.4.2, a system with a rational $H(z)$ is causal iff the order of the numerator is not greater than the order of the denominator. Here, order of the numerator (3) is greater than the order of the denominator (2). Therefore, the system is not causal.

2. Given:
$$H(z) = \frac{1}{1 - \frac{1}{2}z^{-1}} + \frac{1}{1 - 2z^{-1}}, |z| > 2.$$

Is the system causal? Is the system stable?

Solution:

$$H(z) = \frac{(1 - 2z^{-1}) + (1 - \frac{1}{2}z^{-1})}{(1 - 2z^{-1})(1 - \frac{1}{2}z^{-1})}$$

$$= \frac{2 - \frac{5}{2}z^{-1}}{(1 - 2z^{-1})(1 - \frac{1}{2}z^{-1})}$$

163

Multiplying numerator and denominator by z^2,

$$H(z) = \frac{z(2z - \frac{5}{2})}{(z - \frac{1}{2})(z - 2)}, \quad |z| > 2$$

Therefore the poles are located at $z = 1/2$ and $z = 2$. The ROC is indicated in Figure 4.10. Since both the conditions for causality are satisfied, the system is causal. Since the ROC does not include the unit circle, the system is not stable.

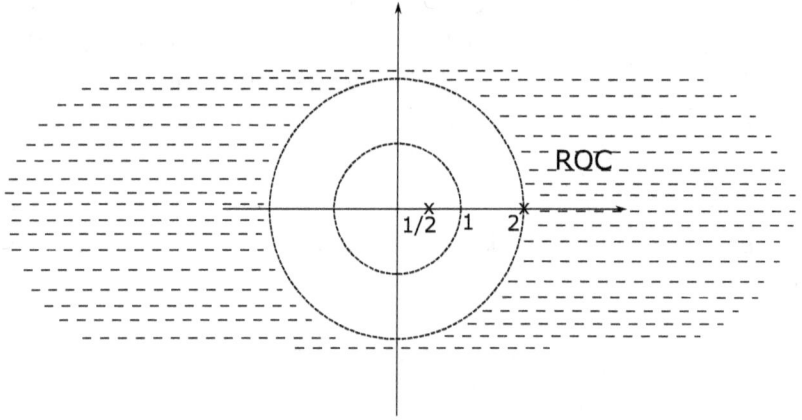

Figure 4.10: ROC: $|z| > 2$

3. Given:

$$H(z) = \frac{1}{1 - \frac{1}{2}z^{-1}} + \frac{1}{1 - 2z^{-1}}, 1/2 < |z| < 2.$$

Is the system causal? Is the system stable?

Solution: From the previous example,

$$H(z) = \frac{z(2z - \frac{5}{2})}{(z - \frac{1}{2})(z - 2)}, \quad 1/2 < |z| < 2$$

The ROC is indicated in Figure 4.11. Since the first condition for causality is not satisfied, the system is not causal. The ROC includes the unit circle, therefore the system is stable.

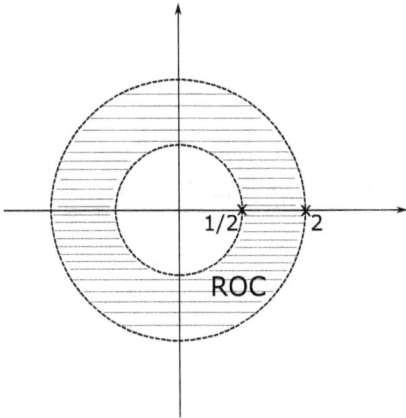

Figure 4.11: ROC: $1/2 < |z| < 2$

4. Given:
$$H(z) = \frac{1}{1 - \frac{1}{2}z^{-1}} + \frac{1}{1 - 2z^{-1}}, |z| < 1/2.$$

Is the system causal? Is the system stable?

Solution: From the previous example,

$$H(z) = \frac{z(2z - \frac{5}{2})}{(z - \frac{1}{2})(z - 2)}, \quad |z| < 1/2$$

The ROC is indicated in Figure 4.12. Since the first condition for causality is not satisfied, the system is not causal. The ROC does not include the unit circle, therefore the system is not stable.

5. Given a LTI system described by: $y[n] - 9/4y[n - 1] + 1/2y[n - 2] = x[n] - 3x[n - 1]$. Specify the ROC of $H(z)$ and determine $h[n]$ for the following conditions: (a) stable system (b) causal system.

Solution: Comparing the difference equation with Equation (4.17),

165

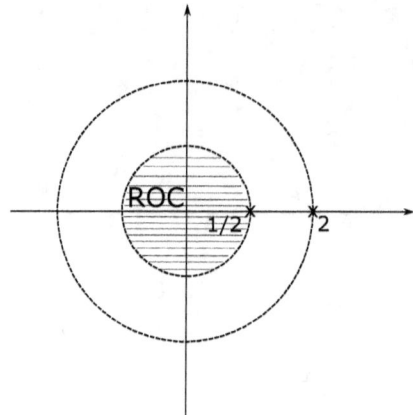

Figure 4.12: ROC: $|z| < 1/2$

$$N = 2 \qquad\qquad M = 1$$
$$a_0 = 1 \qquad\qquad b_0 = 1$$
$$a_1 = -9/4 \qquad\qquad b_1 = -3$$
$$a_2 = 1/2$$

Therefore from Equation (4.19),

$$H(z) = \frac{1 + (-3)z^{-1}}{1 + (-9/4)z^{-1} + (1/2)z^{-2}}$$

Multiplying numerator and denominator by z^2,

$$H(z) = \frac{z(z-3)}{z^2 - 9/4z + 1/2}$$

$$\frac{z(z-3)}{(z-1/4)(z-2)}$$

$$\frac{H(z)}{z} = \frac{(z-3)}{(z-1/4)(z-2)}$$

Using partial fraction expansion,

$$\frac{H(z)}{z} = \frac{(z-3)}{(z-1/4)(z-2)} \qquad\qquad = \frac{A}{z-1/4} + \frac{B}{z-2}$$

Solving for A and B, $A = 11/7$ and $B = -4/7$.

Therefore

$$\frac{H(z)}{z} = \frac{11/7}{z - 1/4} + \frac{-4/7}{z - 2}$$

$$H(z) = \frac{11}{7}\frac{z}{z - 1/4} - \frac{4}{7}\frac{z}{z - 2}$$

$$= \frac{11}{7}\frac{1}{1 - 1/4z^{-1}} - \frac{4}{7}\frac{1}{1 - 2z^{-1}}. \qquad (4.20)$$

(a) <u>For stable system</u>: To determine $h[n]$, we need to know $H(z)$ and the ROC. For the system to be stable, from the first condition in Section 4.4.3, the ROC must include the unit circle. Therefore, the ROC must be $1/4 < |z| < 2$. This is shown in Figure 4.13.

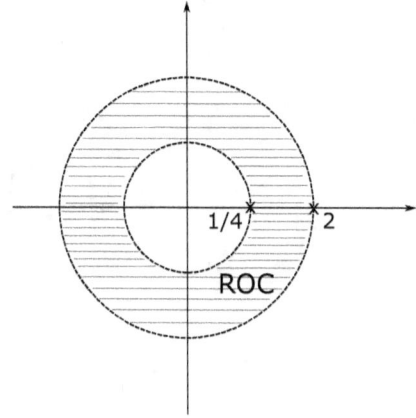

Figure 4.13: ROC: $1/4 < |z| < 2$

To determine $h[n]$ we will utilize the known Z transforms. From Equation (4.20),

$$\frac{11}{7}\frac{1}{1 - 1/4z^{-1}} \xleftrightarrow{\alpha=1/4,|z|>1/4} \frac{11}{7}(\frac{1}{4})^n u[n]$$

$$\frac{4}{7}\frac{1}{1 - 2z^{-1}} \xleftrightarrow{\alpha=2,|z|<2} \frac{4}{7}(-2^n u[-n-1])$$

Therefore,

$$h[n] = \frac{11}{7}\left(\frac{1}{4}\right)^n u[n] + \frac{4}{7}2^n u[-n-1].$$

(b) <u>For causal system</u>: To determine $h[n]$, we need to know $H(z)$ and the ROC. For the system to be causal, from the first condition in Section 4.4.2, the ROC is the exterior of a circle outside the outermost pole. Therefore, the ROC must be $|z| > 2$. This is shown in Figure 4.14.

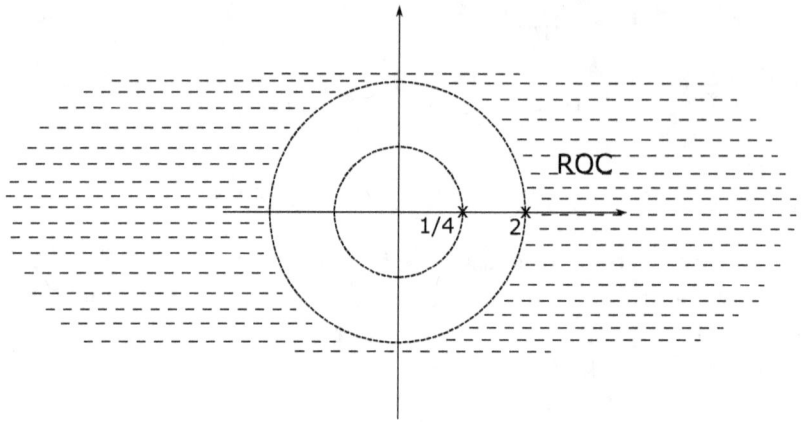

Figure 4.14: ROC: $|z| > 2$

To determine $h[n]$ we will utilize the known Z transforms. From Equation (4.20),

$$\frac{11}{7}\frac{1}{1 - 1/4z^{-1}} \xleftrightarrow{\alpha=1/4,|z|>1/4} \frac{11}{7}(\frac{1}{4})^n u[n]$$

$$\frac{4}{7}\frac{1}{1 - 2z^{-1}} \xleftrightarrow{\alpha=2,|z|>2} \frac{4}{7}2^n u[n]$$

Therefore,

$$h[n] = \frac{11}{7}\left(\frac{1}{4}\right)^n u[n] - \frac{4}{7}2^n u[n].$$

6. Determine the impulse and step response of the following causal system. Discuss on stability. $y[n] - y[n-1] - 2y[n-2] = x[n-1] + 2x[n-2]$.

Solution: Comparing the difference equation with Equation (4.17),

$N = 2$	$M = 2$
$a_0 = 1$	$b_0 = 0$
$a_1 = -1$	$b_1 = 1$
$a_2 = -2$	$b_2 = 2$

Therefore from Equation (4.19),

$$H(z) = \frac{z^{-1} + 2z^{-2}}{1 - z^{-1} - 2z^{-2}}$$

Multiplying numerator and denominator by z^2,

$$H(z) = \frac{z + 2}{z^2 - z - 2}$$
$$= \frac{z + 2}{(z + 1)(z - 2)} \tag{4.21}$$

Using partial fraction expansion,

$$H(z) = \frac{z + 2}{(z + 1)(z - 2)} \qquad = \frac{A}{z + 1} + \frac{B}{z - 2}$$

Solving for A and B, $A = -1/3$ and $B = 4/3$.

Therefore

$$H(z) = \frac{-1}{3} \frac{1}{z + 1} + \frac{4}{3} \frac{1}{z - 2}. \tag{4.22}$$

To determine $h[n]$, we need to know $H(z)$ and the ROC. Since the system is causal, from Section 4.4.2, the ROC must be $|z| > 2$. This is shown in Figure 4.15.

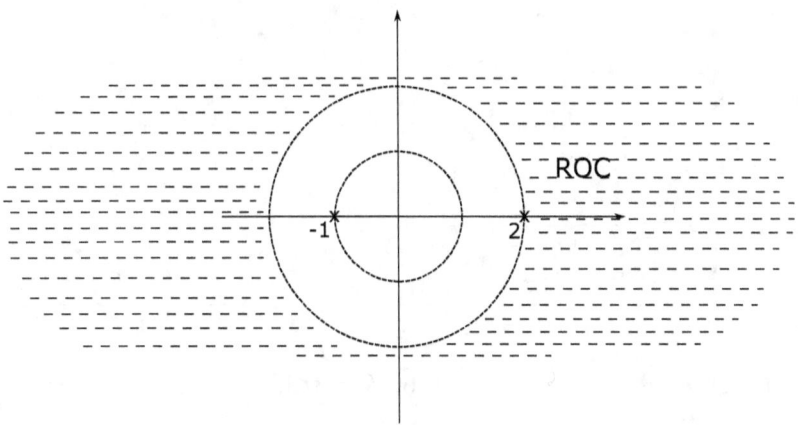

Figure 4.15: ROC: $|z| > 2$

169

Using the known Z transform,

$$a^n u[n] \overset{Z}{\leftrightarrow} \frac{1}{1 - az^{-1}} = \frac{z}{z - a}, |z| > |a|^2$$

and applying the time-shift property,

$$a^{n-1} u[n - 1] \overset{Z}{\leftrightarrow} z^{-1} \frac{z}{z - a}, |z| > |a|, \tag{4.23}$$

the Z transform of the two terms in Equation (4.22) become,

$$\frac{-1}{3} \frac{1}{z + 1} \overset{a=-1, |z|>1}{\longleftrightarrow} \frac{-1}{3} (-1)^{n-1} u[n - 1]$$

$$\frac{4}{3} \frac{1}{z - 2} \overset{a=2, |z|>2}{\longleftrightarrow} \frac{4}{3} (2)^{n-1} u[n - 1]$$

Therefore Equation (4.22) becomes,

$$h[n] = \frac{-1}{3} (-1)^{n-1} u[n - 1] + \frac{4}{3} (2)^{n-1} u[n - 1].$$

Since the ROC does not include the unit circle, the system is not stable (Section 4.4.3).

To calculate the step response ($y[n]$), we first calculate $Y(z) = H(z)X(z)$, where $H(z)$ is given by Equation (4.21) and $x[n] = u[n]$.

The Z transform of $x[n]$ is given by

$$x[n] = u[n] \overset{Z}{\leftrightarrow} X(z) = \frac{z}{z - 1}, |z| > 1.$$

Therefore,

$$Y(z) = H(z)X(z), |z| > 2$$

$$= \frac{z + 2}{(z + 1)(z - 2)} \frac{z}{z - 1}$$

$$= \frac{z(z + 2)}{(z + 1)(z - 2)(z - 1)} \qquad = \frac{A}{z + 1} + \frac{B}{z - 2} + \frac{C}{z - 1}$$

Solving for A, B, C gives $A = -1/6, B = 8/3, C = -3/2$.

---

[2]Note that the known Z transform $-a^n u[-n - 1] \overset{Z}{\leftrightarrow} \frac{z}{z-a}, |z| < |a|$ is not being used, since it will not result in the overall ROC $|z| > 2$.

Therefore,

$$Y(z) = \frac{-1}{6}\frac{1}{z+1} + \frac{8}{3}\frac{1}{z-2} - \frac{3}{2}\frac{1}{z-1}, |z| > 2.$$

Utilizing Equation (4.23),

$$y[n] = \frac{-1}{6}(-1)^{n-1}u[n-1] + \frac{8}{3}(2)^{n-1}u[n-1] - \frac{3}{2}u[n-1].$$

5 Sampling

The goal of any communication system is to transmit a signal (via a transmitter), and at the receiver, to reconstruct a signal which closely represents the orginal transmitted signal. To do so, a continuous time signal is first represented by its samples at the transmitter side and the receiver must reconstruct the signal from these transmitted samples. In general, we can reconstruct an infinite number of signals from a given set of samples. The Sampling Theorem helps to uniquely reconstruct the original signal from the given set of samples. Thus, this chapter focuses on the concepts of sampling, Sampling Theorem and aliasing. The concept map for the topics covered in this chapter is shown in Figure 5.1.

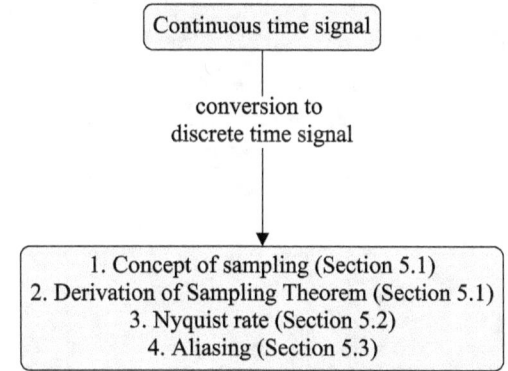

Figure 5.1: Concept map of topics in Chapter 5

5.1 Derivation of Sampling Theorem

The sampling theorem aims to answer the following question: At what frequency should we sample a continuous time signal such that this signal can be reconstructed from its samples?

Consider a signal $x(t)$ as shown in Figure 5.2.

To sample $x(t)$ we multiply $x(t)$ with the impulse train or sampling function $p(t)$ (shown in Figure 5.3).

173

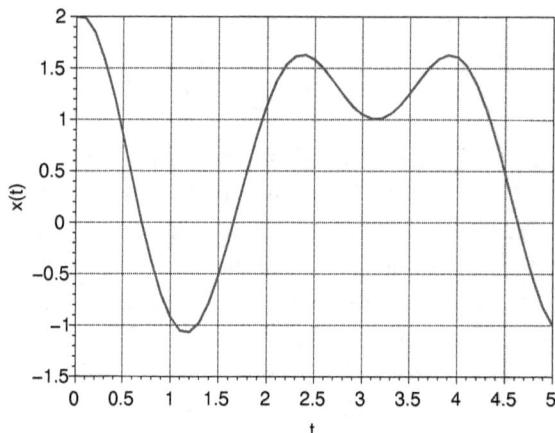

Figure 5.2: A continuous time signal $x(t)$

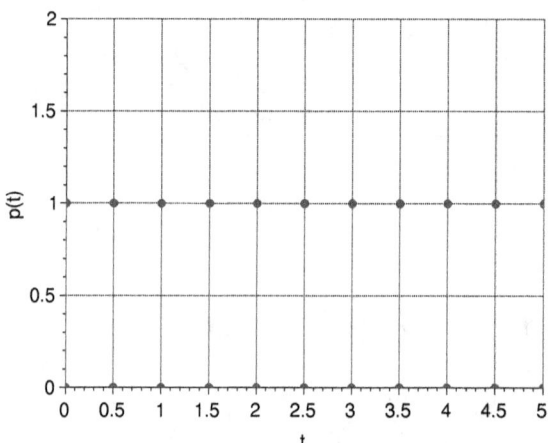

Figure 5.3: Impulse train $p(t)$

Mathematically, $p(t)$ is given by:

$$p(t) = \sum_{n=-\infty}^{\infty} \delta(t - nT) \tag{5.1}$$

Note that $p(t)$ is a periodic function that is periodic in T. T is called the sampling period. The sampling frequency ω_s is given by $\omega_s = \frac{2\pi}{T}$.

After sampling $x(t)$, the sampled signal $x_p(t)$ is shown in Figure 5.4. Note that

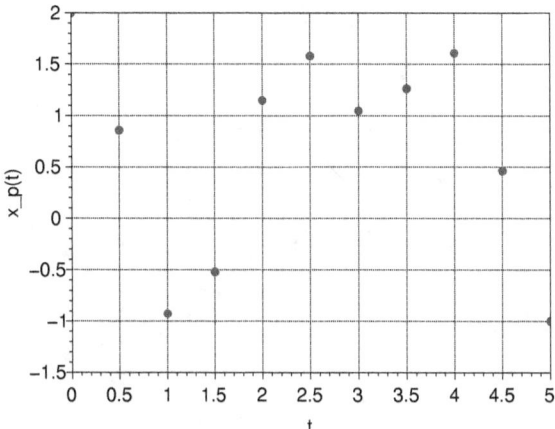

Figure 5.4: The sampled signal $x_p(t)$

if the sampling period T is lower, then $x(t)$ is sampled more finely, and $x_p(t)$ will more closely approximate $x(t)$. This is shown in Figure 5.5.

Mathematically, $x_p(t) = x(t) \times p(t)$.

The Fourier Transform of $x_p(t)$, given by $X_p(j\omega)$, is:

$$X_p(j\omega) = X(j\omega) * P(j\omega) \tag{5.2}$$

$$= \frac{1}{2\pi} \int_{-\infty}^{\infty} X(j\theta) P(j(\omega - \theta)) d\theta \tag{5.3}$$

Let us assume $X(j\omega)$ is as shown in Figure 5.6. Note that the ω_M is the maximum frequency of $X(j\omega)$.

$p(t)$ is given by Equation (5.1). Recall that the Fourier Transform of the impulse train function is also an impulse train. The Fourier Transform of $p(t)$, denoted by $P(j\omega)$ (shown in Figure 5.7), is given by:

$$P(j\omega) = \frac{2\pi}{T} \sum_{k=-\infty}^{\infty} \delta(\omega - k\omega_s) \tag{5.4}$$

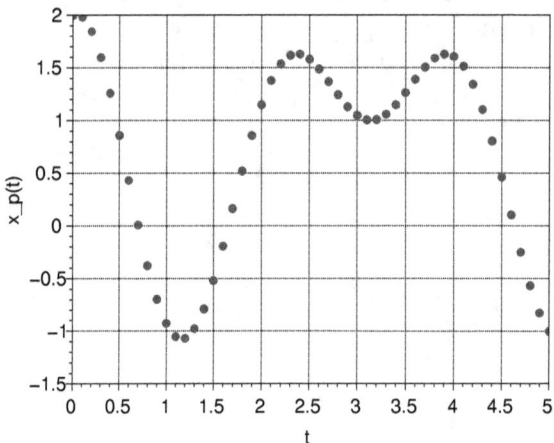

Figure 5.5: The sampled signal $x_p(t)$ with lower T (i.e. higher sampling frequency)

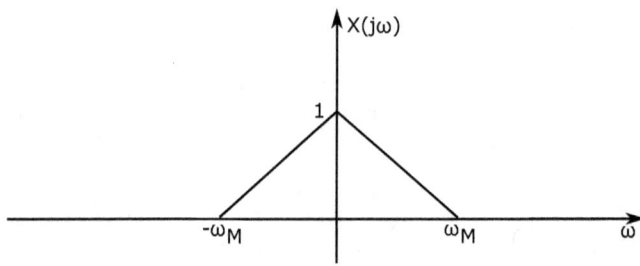

Figure 5.6: $X(j\omega)$

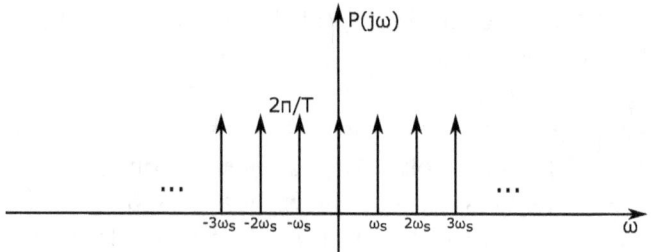

Figure 5.7: $P(j\omega)$

Substituting Equation (5.4) in Equation (5.3):

$$X_p(j\omega) = \frac{1}{2\pi} \int_{-\infty}^{\infty} X(j\theta) \frac{2\pi}{T} \sum_{k=-\infty}^{\infty} \delta(\omega - k\omega_s - \theta) d\theta \tag{5.5}$$

$$= \frac{1}{T} \sum_{k=-\infty}^{\infty} \int_{-\infty}^{\infty} X(j\theta)\delta(\omega - k\omega_s - \theta) d\theta \tag{5.6}$$

From the sampling property of the impulse function

$$\int_{-\infty}^{\infty} X(j\theta)\delta(\omega - k\omega_s - \theta) d\theta = X(j(\omega - k\omega_s)) \tag{5.7}$$

Therefore, Equation (5.6) becomes

$$X_p(j\omega) = \frac{1}{T} \sum_{k=-\infty}^{\infty} X(j(\omega - k\omega_s)) \tag{5.8}$$

The plot of $X_p(j\omega)$ is shown in Figure 5.8. Note that $X_p(j\omega)$ is a periodic function

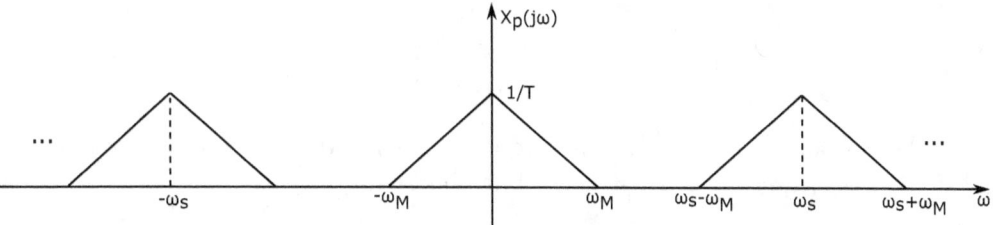

Figure 5.8: $X_p(j\omega)$

in ω_s, and it consists of shifted replicas of $X(j\omega)$. Recall that T or ω_s is our design parameter.

177

Recall that the original signal is $x(t)$ and the sampled signal is $x_p(t)$. Remember that we would like to reconstruct $x(t)$ from $x_p(t)$. We will be achieving this by looking at the *frequency domain* of $x_p(t)$.

If we were to send $x_p(t)$ through an ideal low-pass filter (whose frequency response is shown in Figure 5.9) with a cutoff $\omega_c > \omega_M$, then the output signal of the low-pass filter will have a frequency response as shown in Figure 5.10 (recall: convolution in the time domain leads to multiplication in the frequency domain).

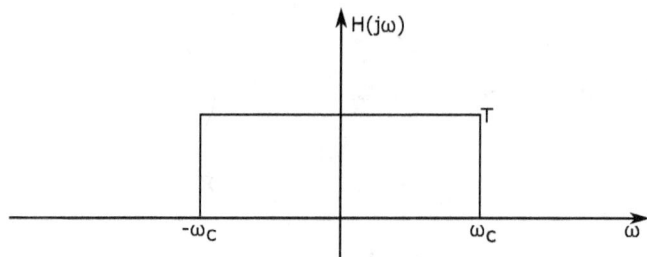

Figure 5.9: Frequency response of an ideal low-pass filter

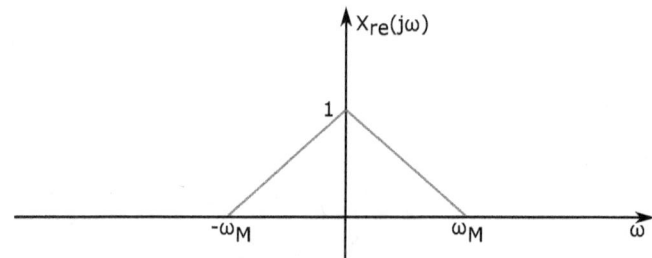

Figure 5.10: Frequency response of the reconstructed signal (after low-pass filter)

This reconstruction is possible when

$$\omega_s - \omega_M > \omega_M \tag{5.9}$$

$$\omega_s > 2\omega_M \tag{5.10}$$

This relation between the sampling frequency (ω_s) and the maximum frequency of the signal $x(t)$ (ω_M) gives us the sampling theorem.

5.2 Sampling Theorem

Let $x(t)$ be a bandlimited signal with $X(j\omega) = 0$ for $|\omega| > \omega_M$. Then $x(t)$ is uniquely determined by its samples $x(nT), n = 0, \pm 1, \pm 2, \ldots$ if $\omega_S > 2\omega_M$, where $\omega_S = 2\pi/T$.

Given these samples we can reconstruct $x(t)$ by passing the sampled signal through an ideal low pass filter with gain T and cutoff frequency greater than ω_M and less than $\omega_S - \omega_M$.

When the sampling frequency $\omega_s = \omega_{NR} = 2\omega_M$, that is the minimum sampling frequency, it is called the Nyquist Rate.

5.2.1 Solved Examples

1. Determine the Nyquist rate of sampling of the signal $x(t) = 10\,\text{sinc}(500t)$.

Solution:

$$sinc(t) = \frac{\sin(\pi\theta)}{\pi\theta}$$

$$sinc(500t) = \frac{\sin(500\pi t)}{500\pi t}$$

$$\Rightarrow \omega_M = 500\pi$$

$$\Rightarrow \omega_{NR} = 2\omega_M = 1000\pi \ \text{rad/s}$$

$$\Rightarrow f_{NR} = \frac{\omega_{NR}}{2\pi} = 500 \ \text{samples/sec.}$$

2. Determine the Nyquist rate of sampling of the signal $x(t) = 2\sin(250\pi t) + 3\cos^2(500t)$.

Solution:

$$x(t) = 2\sin(250\pi t) + 3\frac{1 + \cos(1000t)}{2}$$

$$\Rightarrow \omega_M = 1000 \ \text{rad/s}$$

$$\Rightarrow \omega_{NR} = 2\omega_M = 2000 \ \text{rad/s.}$$

3. Determine the condition on the sampling interval T_s so that the following signal can be reconstructed from its samples: $x(t) = \cos(\pi t) + 3\sin(2\pi t) + \sin(4\pi t)$.

Solution:

$$\omega_M = 4\pi$$
$$\Rightarrow \omega_s > 2\omega_M$$
$$\Rightarrow \omega_s > 8\pi \ \text{rad/s}$$
$$\Rightarrow \frac{2\pi}{T_s} > 8\pi$$
$$\Rightarrow T_s < \frac{1}{4} \ \text{sec.}$$

---

4. Determine the condition on the sampling interval T_s so that each of the following signal can be reconstructed from its samples: $x(t) = \cos(2\pi t)\frac{\sin(\pi t)}{\pi t} + 3\sin(6\pi t)\frac{\sin(2\pi t)}{\pi t}$.

Solution: Using the relations:

$$\cos(A)\sin(B) = \frac{1}{2}[\sin(A+B) - \sin(A-B)]$$
$$\sin(A)\sin(B) = \frac{1}{2}[\cos(A-B) - \cos(A+B)]$$

we can write

$$x(t) = \frac{1}{\pi t}\frac{[\sin(3\pi t) - \sin(\pi t)]}{2} + \frac{3}{\pi t}\frac{[\cos(4\pi t) - \cos(8\pi t)]}{2}$$
$$= \frac{\sin(3\pi t)}{2\pi t} - \frac{\sin(\pi t)}{2\pi t} + \frac{3\cos(4\pi t)}{2\pi t} - \frac{3\cos(8\pi t)}{2\pi t}$$
$$\Rightarrow \omega_M = 8\pi$$
$$\omega_s > 2\omega_M$$
$$\Rightarrow \omega_s > 2(8\pi) \ \text{rad/s}$$
$$\Rightarrow \frac{2\pi}{T_s} > 2(8\pi)$$
$$\Rightarrow T_s < \frac{1}{8} \ \text{sec.}$$

5.3 Aliasing

When $\omega_S < 2\omega_M$ (i.e. when we under sample), the plot of $X_p(j\omega)$ is shown in Figure 5.11. In this case, $X(j\omega)$ (Figure 5.6) is no longer replicated in $X_p(j\omega)$ and therefore we cannot reconstruct or recover $X(j\omega)$ from $X_p(j\omega)$ using a low pass filter. This effect is known as aliasing.

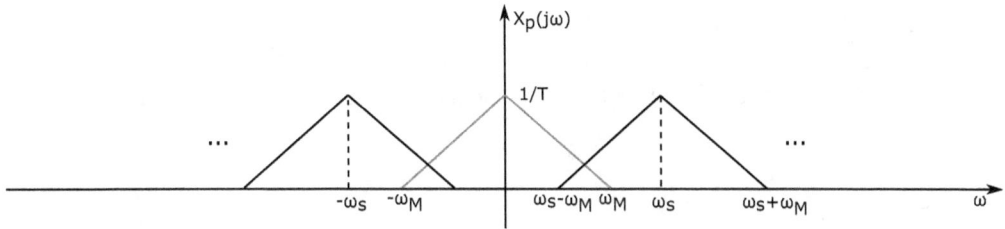

Figure 5.11: Effect of aliasing

5.3.1 Solved Example

1. Consider the continuous time band-limited signal $x(t)$ with a spectrum $X(f)$ as shown in Figure 5.12. Sketch the spectrum of the discrete time signal $x_1[n]$ and $x_2[n]$ obtained from $x(t)$ by sampling at 5 kHz and 3 kHz respectively.

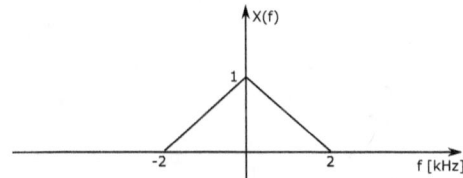

Figure 5.12: Spectrum $X(f)$

Solution: $f_M = 2$ kHz.

When $x(t)$ is sampled at 5 kHz:

$f_s = 5 > 2f_M = 4$ (Nyquist rate). Therefore there will be no overalp in the frequency spectrum. The spectrum of the sampled signal is shown in Figure 5.13.

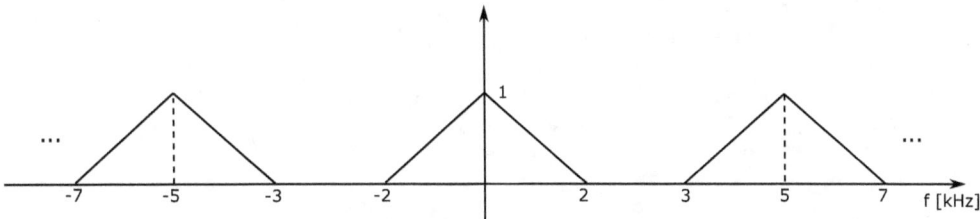

Figure 5.13: Spectrum when sampled at 5 kHz

181

When $x(t)$ is sampled at 3 kHz:

$f_s = 3 < 2f_M = 4$ (Nyquist rate). Therefore there will be overalp in the frequency spectrum. The spectrum of the sampled signal is shown in Figure 5.14.

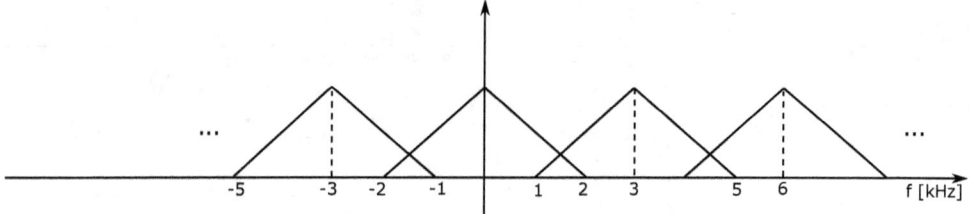

Figure 5.14: Spectrum when sampled at 3 kHz

Bibliography

[1] A. V. Oppenheim, A. S. Willsky, and S. H. Nawab, *Signals and Systems*, 2nd ed. Pearson Education Ltd., 2013.

[2] S. Haykin and B. Van Veen, *Signals and Systems*, 2nd ed. Wiley, 2002.

[3] P. R. Rao and S. Prakriya, *Signals and Systems*, 2nd ed. McGraw Hill Education (India), 2013.

[4] A. A. Kumar, *Signals and Systems*, 3rd ed. PHI Learning, 2013.

[5] H. Hsu, *Schaum's Outline of Signals and Systems*, 2nd ed. McGraw-Hill Education, 2010.

About the Author

Krishna Kumar Kishor received his Bachelor, Master and Ph.D. degrees in Electrical Engineering from the University of Toronto, Canada. In his graduate studies he specialized in Electromagnetics with a focus on antenna design. He is a faculty in the Department of Electronics and Communication Engineering at Ahalia School of Engineering and Technology, Palakkad, Kerala, India. He is a Senior Member of the IEEE, IEEE Antennas and Propagation Society (IEEE - APS), IEEE Microwave Theory and Techniques Society (IEEE - MTTS), an Associate Member of the Institute of Engineers India (IEI) and a Life Member of the Indian Society for Technical Education (ISTE).

www.ingramcontent.com/pod-product-compliance
Lightning Source LLC
Chambersburg PA
CBHW052034280526
45791CB00010B/2963